Springer Asia Pacific Mathematics Series

Volume 10

The Springer Asia Pacific Mathematics Series promotes high-quality scientific research connected to Asia-Pacific countries. Monographs, edited volumes, lecture notes, and textbooks will cover topics related to pure and applied mathematics as well as interdisciplinary research.

***

All material will be rigorously peer-reviewed and should be related to mathematical research conducted at or with institutions in the APAC region. This series is open to subseries, which may be topical or exclusively supplied by a single research institution.

***

Shuichi Yukita

First Lectures in Algebra

Why do Normal Subgroups and Ideals Matter?

Springer

Shuichi Yukita
Hosei University
Tokyo, Japan

ISSN 3091-2555 ISSN 3091-2563 (electronic)
Springer Asia Pacific Mathematics Series
ISBN 978-981-95-5456-0 ISBN 978-981-95-5457-7 (eBook)
https://doi.org/10.1007/978-981-95-5457-7

Mathematics Subject Classification: 00A09, 11-01, 13-01, 00A06, 12-01, 20-01, 16-01

This Springer imprint is published by the registered company Springer Nature Singapore Pte Ltd.
The registered company address is: 152 Beach Road, #21-01/04 Gateway East, Singapore 189721, Singapore

Preface

It was probably in the 1970s that linear algebra became a compulsory subject in the curricula of engineering departments. Nobody argues today about its key position in college mathematics, whereas theories of groups, rings, and fields are still elective in most curricula.

Applications requiring knowledge of groups, rings, and fields are growing in numbers. Fully established but still developing areas are *coding* and *cryptography*. In addition to these fields, we should mention emerging applications in *deep learning* with *tensor algebra* nowadays. Algebraic geometry, the geometry of *algebraic varieties*, still being a champion of abstraction, is used concretely in those emerging application areas, where familiarity with groups, rings, and fields is required.

This book is for science and engineering students rather than mathematics majors, yet there is no cutting of corners in the argumentation. The book is, so to say, a rigorous but gentle introduction to algebra. Even in mathematics department classes, undergraduates often have questions like why normal subgroups and ideals matter. Such students will find standard answers with full intuition and rigor in this book. The previews are:

- A *normal subgroup* is the kernel of a group homomorphism from the parent group to some other group. A normal subgroup yields a factor group of the parent. A subgroup that is not normal has neither of these properties.
- An *ideal*, a special kind of additive subgroup of a ring, is the kernel of a ring homomorphism from the parent ring to some other ring. An ideal yields a factor ring of the parent. An additive subgroup of a ring that is not an ideal has neither of these properties.

The reader will fully appreciate the beauty of the parallelism between these answers after reading Chaps. 2–10.

The major features are as follows.

- Chapter 1 has a hidden theme. The Euclidean algorithm is presented in an elementary manner. However, some readers may soon find out that the presentation

is strongly conscious of the theory of modules over Euclidean domains. The author believes that this approach is the most lucid and easiest one even in a high-school math class.
- The concepts of *orbit* and *cluster* are recurring. Orbits of a group action are shown with intuitive drawings. The reader will be led to notice that the structures of normal subgroups and ideals are fully understood operationally in the context of homomorphisms. We emphasize *experience in drawing and computing* of orbits and homomorphisms.
- We place emphasis on the *design patterns* of proofs. Design patterns are used in various areas such as architecture and software engineering. We borrow ideas from these other disciplines.
- Many drawings will appear. These are vital constituents of the book for the *concept visualization*. When tackling quizzes and exercises, the reader is advised first to draw a descriptive picture or diagram, then calculate with some concrete data, then design the flow of reasoning, and finally write out a proof.
- Answers to quizzes and exercises are fully included. On first reading, the reader may jump to the answer after 10 minutes of struggle on one's own.

The following topics are omitted, which are often included in many introductory books.

- Sylow's theorems. A collection of theorems that penetrate into the structure of finite groups. Using Sylow's theorems, some of the exercises in this book would have easier and more elegant solutions.
- The fundamental theorem of finitely generated abelian groups. It is a surprisingly simple and beautiful theorem, but it requires a lot of preparation.
- The Jordan-Hölder theorem for composition series. This requires further preparation.

All of the items require a considerable number of pages. I decided to keep this book concise while sticking to the style of *concept visualization*. The reader will get enough basic skills to further study algebra with standard textbooks.

Chapter 12 presents a gentle introduction to error-correcting codes as an application of finite fields. Everyone who has read Chap. 11 will definitely enjoy it.

The reader is assumed to have basic knowledge of high-school algebra and to be familiar with linear algebra of college first-year level.

This book emerged from a series of discussions in the 1980s with Mitsuo Ohno, who was a math enthusiast and operated an academic publishing company in Tokyo. I was then a PhD candidate who had an opportunity to deliver lectures in algebra at an engineering department. Ohno and I talked over the contents and styles of math lectures for engineers or math enthusiasts. Many ideas were developed through these talks.

I would like to thank Professor Kenji Omori for reviewing not only the mathematical content but also the narrative style. I would also like to thank Daiki

Sato for his advice on the structure of this book from a reader-first perspective. I would like to thank my wife Minako for everything.

I dedicate this book to the late Mr. Ohno.

Tokyo, Japan
October 2025

Shuichi Yukita

Contents

Chapter 1
Preliminaries

This chapter collects information that is necessary or useful when reading this book, from high-school mathematics to undergraduate mathematics. Let us look at sets and mappings, prime factorization of natural numbers, remainder and factor theorems for polynomials, and useful matrix calculation techniques.

1.1 Sets

In this book, it is sufficient to simply think of a set as a collection of things. Namely, we adopt naive set theory. Commonly used symbols for sets and mappings are used, but some symbols require caution since there are several styles in use. Therefore, we have to settle our convention on how to use symbols in this book.

When x is an element of a set A, we write $x \in A$. When A is a subset of B, we write $A \subset B$. In this case, we do not exclude the possibility that $A = B$. Instead of the symbol $\subset$, symbols $\subseteq$ and $\subseteqq$ are often used in many books and papers, but in this book, $A \subset B$ is used. When A is a proper subset of B, if we need to emphasize $A \neq B$, we write $A \subsetneqq B$.

A set can be defined by enumerating the elements that belong to it. For infinite sets, it is awkward to take this approach. However, for example, all integers greater than or equal to 0 can be expressed as

$$\{0, 1, 2, 3, \cdots\}, \tag{1.1}$$

which gives no room for misunderstanding.

We use standard notation for frequently appearing sets. The set of all integers, rational, real, and complex numbers are denoted by $\mathbb{Z}$, $\mathbb{Q}$, $\mathbb{R}$, and $\mathbb{C}$, respectively. Then, Eq. (1.1) can be expressed as

S. Yukita, *First Lectures in Algebra*, Springer Asia Pacific Mathematics Series 10,
https://doi.org/10.1007/978-981-95-5457-7_1

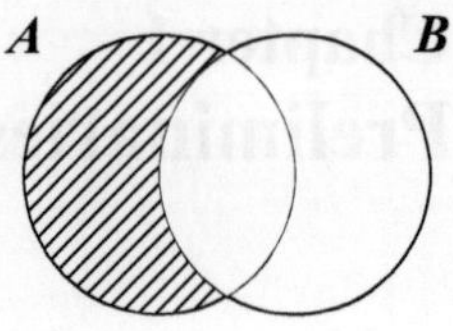

Fig. 1.1 $A \setminus B$

$$\{n \in \mathbb{Z} \mid n \geq 0\}.$$

The union of sets A and B is denoted by $A \cup B$. The intersection by $A \cap B$. The difference is denoted by $A \setminus B$ or $A - B$, more specifically

$$A \setminus B = \{x \in A \mid x \notin B\}.$$

See Fig. 1.1.

If $A \cap B = \emptyset$, the union $A \cup B$ is called the *disjoint union* of A and B and often denoted by $A \sqcup B$.

The product (Cartesian product) of sets A and B is denoted by $A \times B$. The definition is as follows:

$$A \times B = \{(a, b) \mid a \in A, b \in B\}.$$

Note that (a, b) denotes the ordered pair of a and b. So, we have $(a, b) \neq (b, a)$ if $a \neq b$. The unordered pair may be expressed as $\{a, b\}$ using the set notation that guarantees $\{a, b\} = \{b, a\}$.

The set of all real numbers $\mathbb{R}$ can be identified with a straight line. The product $\mathbb{R} \times \mathbb{R}$ can be identified with a plane, which is often expressed as $\mathbb{R}^2$. By iteration of the product, we have $\mathbb{R}^n$ and $\mathbb{C}^n$ for all natural numbers n. To be more specific, the latter is defined as

$$\mathbb{C}^n = \{(z_1, z_2, \dots, z_n) \mid z_i \in \mathbb{C}\}.$$

1.2 Mappings

The concept of mappings or functions contains some pitfalls. The terms *mapping*, *map* (used as a noun), and *function* mean completely the same thing.

Let $f : A \to B$ be a map from A to B. We say f is an *injection* if $f(x) \neq f(y)$ for all $x, y \in A$ with $x \neq y$. We say f is a *surjection* if for all $z \in B$ there is an element $x \in A$ such that $f(x) = z$. Such an x may, of course, depend on z. We say f is a *bijection* if it is both injective and surjective. Adjectives *injective*, *surjective*, and *bijective* are used in an obvious manner in context.

Example 1.1 Let $A = \{1, 2, 3\}$, $B = \{1, 2, 3, 4\}$. We define $f_1 : A \to B$ as follows:

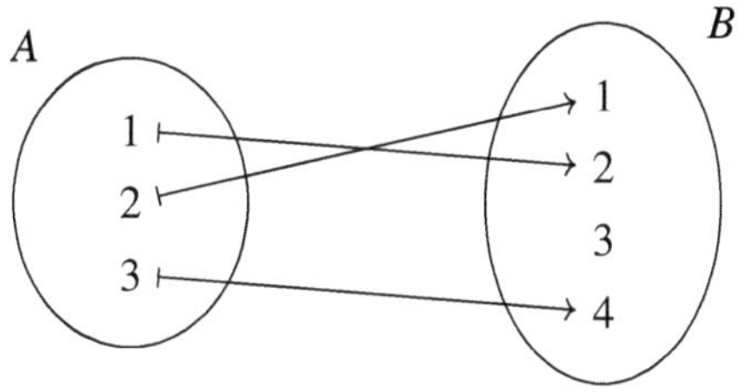

No two elements are sent to the same destination. Thus, f_1 is an injection.

We define $f_2 : A \to B$ as follows:

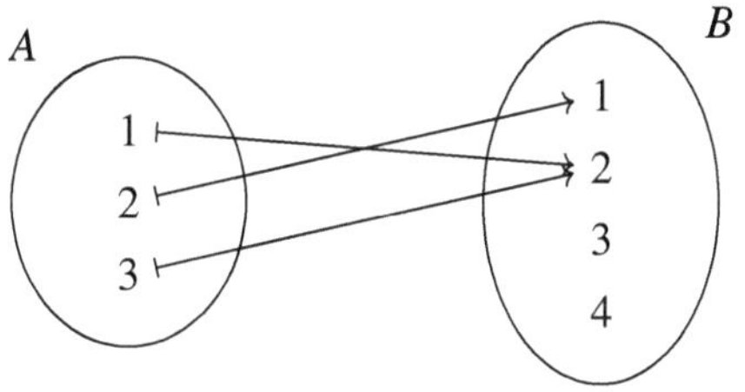

Two elements 1 and 3 are sent to the same destination in B. So, f is not an injection.

The following correspondence is not a mapping from A to B:

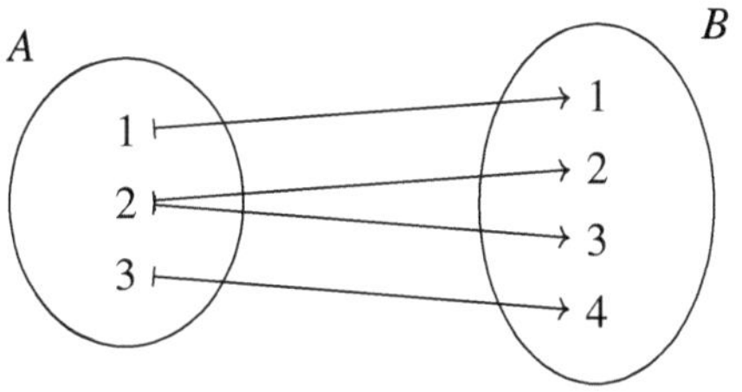

To be a mapping, exactly one arrow must start from every element.

Though there are no surjections from A to B, we have several from B to A. We define $f_3 : B \to A$ as follows:

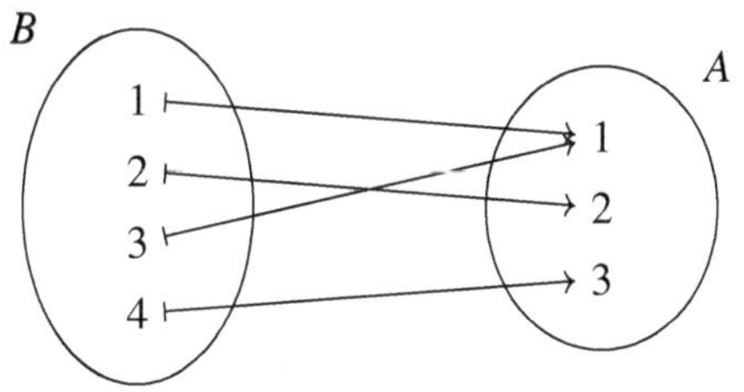

The mapping $f_3 : B \to A$ is a surjection.

Remark 1.2 We should not mix up arrows $\to$ and $\mapsto$. The arrow $a \mapsto b$ means an elementwise correspondence, while the arrow $A \to B$ can be seen as the whole data of elementwise correspondence.

1.3 Natural Numbers and Primes

The reader should be familiar with the fact that any natural number is uniquely decomposed as a product of prime numbers. However, most people other than mathematics majors have little experience in exact reasoning. As a preparation to rigorous arguments in later chapters, let us discuss natural numbers and primes in detail.

We first review the concepts of quotient and remainder for integer division.

Theorem 1.3 (Euclidean Division) *Given an integer x and a natural number $n \geq 2$, there exist unique integers x and $0 \leq r < n$ that satisfy*

$$x = qn + r.$$

So obtained integers q and r are called the *quotient* and *remainder* of the division of x by n. Especially when $r = 0$, we say that x divides n.

For example, take $x = -15$ and $n = 7$, then we have $q = -3$ and $r = 6$ as follows:

$$-15 = -3 \times 7 + 6.$$

Let us next review the concept of prime numbers, or just primes.

Definition 1.4 A *prime number* or *prime* is a natural number greater than 1 that is not a product of two smaller natural numbers.

Proposition 1.5 *Any natural number can be expressed as a product of prime numbers, where 1 and primes themselves are considered as a product of primes.*

Proof We prove by contradiction. We suppose there is a natural number that cannot be expressed as a product of primes. Let n be the smallest of all such natural numbers. Since n is neither 1 nor a prime, n can be expressed as a product $n = ab$ with $1 < a < n$ and $1 < b < n$. By the minimality of n, natural numbers a and b should both be expressed as a product of primes, consequently so is n, which leads to a contradiction. □

We have just shown that decomposition to primes is always possible. To show uniqueness, we have to do some preparations.

If d is the greatest common divisor of natural numbers a and b, we write $\gcd(a, b) = d$. We consider the set

$$I = \mathbb{Z}a + \mathbb{Z}b \stackrel{\text{def}}{=} \{xa + yb \mid x, y \in \mathbb{Z}\}.$$

The set I is closed under addition. In fact, for any two elements of I, say $x_1a + y_1b$ and $x_2a + y_2b$, we have

$$(x_1a + y_1b) + (x_2a + y_2b) = (x_1 + x_2)a + (y_1 + y_2)b,$$

which is an element of I.

Let $\gcd(a, b) = d$. There are natural numbers a_1 and b_1 such that $a = a_1d$ and $b = b_1d$. Then, we can rewrite $xa + yb$ as

$$xa + yb = (xa_1 + yb_1)d.$$

Let J be the subset of $\mathbb{Z}$ defined by

$$J = \mathbb{Z}d \stackrel{\text{def}}{=} \{zd \mid z \in \mathbb{Z}\}.$$

Clearly, we have $I \subset J$.

We will show the famous fact $I = J$ using the Euclidean algorithm.

Theorem 1.6 (GCD Generates the Whole) *Let $d = \gcd(a, b)$ for natural numbers a and b. If we set*

$$I = \mathbb{Z}a + \mathbb{Z}b,$$
$$J = \mathbb{Z}d,$$

then we have $I = J$.

Proof Let us write $a = a_1$ and $b = a_2$. We can assume $a_1 \geq a_2$ without losing generality.

If a_2 divides a_1, then we have $d = a_2$ and that there exists a natural number q with $a_1 = qd$. Thus, we have the following:

$$\begin{aligned} I = \mathbb{Z}a_1 + \mathbb{Z}a_2 &= \{xqd + yd \mid x, y \in \mathbb{Z}\} \\ &= \{(xq + y)d \mid x, y \in \mathbb{Z}\}. \end{aligned}$$

Since $xq + y$ varies over all integers, we see that $I = J$.

Quiz 1.1 Show that $xq + y$ can take any values from $\mathbb{Z}$ when x and y vary over $\mathbb{Z}$.

If a_2 does not divide a_1, then we consider the integer division of a_1 by a_2 as

$$a_1 = q_1 a_2 + a_3, \tag{1.2}$$

where q_1 is the quotient and a_3 is the remainder with $(a_1 >)\, a_2 > a_3$. The Euclidean algorithm is based on the fact that

$$\gcd(a_1, a_2) = \gcd(a_2, a_3).$$

Let us see if this is true. If k is a common divisor of a_1 and a_2, namely k divides both a_1 and a_2, then by (1.2) we have

$$a_3 = a_1 - q_1 a_2, \tag{1.3}$$

which shows that k divides a_3, therefore k is a common divisor of a_2 and a_3.

Conversely, assume k divides both a_2 and a_3. Equation (1.2) shows that k divides a_1. Thus, k is a common divisor of a_1 and a_2.

We have just discovered that the set of all common divisors of a_1 and a_2 is identical to that of a_2 and a_3. Therefore, we have

$$\gcd(a_1, a_2) = \gcd(a_2, a_3).$$

Equations (1.2) and (1.3) provide more information:

$$\{xa_1 + ya_2 \mid x, y \in \mathbb{Z}\} = \{sa_2 + ta_3 \mid s, t \in \mathbb{Z}\},$$

or concisely

$$\mathbb{Z}a_1 + \mathbb{Z}a_2 = \mathbb{Z}a_2 + \mathbb{Z}a_3. \tag{1.4}$$

Let us prove Eq. (1.4). Using Eq. (1.2), we have

$$\begin{aligned} xa_1 + ya_2 &= x(q_1 a_2 + a_3) + ya_2 \\ &= (xq_1 + y)a_2 + xa_3. \end{aligned}$$

Equating with $sa_2 + ta_3$, we have, in a matrix form,

$$\begin{pmatrix} x & y \end{pmatrix} \begin{pmatrix} a_1 \\ a_2 \end{pmatrix} = \begin{pmatrix} xq_1 + y & x \end{pmatrix} \begin{pmatrix} a_2 \\ a_3 \end{pmatrix} = \begin{pmatrix} s & t \end{pmatrix} \begin{pmatrix} a_2 \\ a_3 \end{pmatrix}.$$

Since

$$\begin{pmatrix} xq_1 + y & x \end{pmatrix} = \begin{pmatrix} x & y \end{pmatrix} \begin{pmatrix} q_1 & 1 \\ 1 & 0 \end{pmatrix},$$

we see that $\begin{pmatrix} x & y \end{pmatrix}$ and $\begin{pmatrix} s & t \end{pmatrix}$ are related as follows:

$$\begin{pmatrix} x & y \end{pmatrix} \begin{pmatrix} q_1 & 1 \\ 1 & 0 \end{pmatrix} = \begin{pmatrix} s & t \end{pmatrix},$$

$$\begin{pmatrix} x & y \end{pmatrix} = \begin{pmatrix} s & t \end{pmatrix} \begin{pmatrix} 0 & 1 \\ 1 & -q_1 \end{pmatrix}.$$

These equations show that for any $x, y \in \mathbb{Z}$ we can find $s, t \in \mathbb{Z}$ such that $xa_1 + ya_2 = sa_2 + ta_3$. Conversely, for any $s, t \in \mathbb{Z}$ we can find $x, y \in \mathbb{Z}$ such that $xa_1 + ya_2 = sa_2 + ta_3$. Now, we conclude that Eq. (1.4) holds.

According to the Euclidean algorithm, we try then to divide a_2 by a_3. We have

$$a_2 = q_2 a_3 + a_4,$$

where q_2 and a_4 are the quotient and remainder, respectively. Repeating the same discussion as above, we have

$$\mathbb{Z}a_2 + \mathbb{Z}a_3 = \mathbb{Z}a_3 + \mathbb{Z}a_4.$$

Iterating this process while a_{k+1} does not divide a_k, we have

$$a_k = q_k a_{k+1} + a_{k+2}, \quad a_k > a_{k+1} > a_{k+2} > 0,$$
$$\mathbb{Z}a_k + \mathbb{Z}a_{k+1} = \mathbb{Z}a_{k+1} + \mathbb{Z}a_{k+2}.$$

The strictly decreasing sequence $\{a_k\}$ of positive integers must end within finite iterations. Let n be the number that satisfies $a_{n+1} > 0$ and $a_{n+2} = 0$. We must have

$$a_n = q_n a_{n+1}$$
$$\mathbb{Z}a_n + \mathbb{Z}a_{n+1} = \mathbb{Z}a_{n+1}.$$

To sum up, we have the following:

$$\mathbb{Z}a_1 + \mathbb{Z}a_2 = \mathbb{Z}a_2 + \mathbb{Z}a_3 = \ldots = \mathbb{Z}a_n + \mathbb{Z}a_{n+1} = \mathbb{Z}a_{n+1},$$
$$\gcd(a_1, a_2) = \gcd(a_2, a_3) = \ldots = \gcd(a_n, a_{n+1}) = a_{n+1} = d.$$

We have established $I = J$. □

Example 1.7 Take $a = 42$ and $b = 30$. We have

$$42\mathbb{Z} + 30\mathbb{Z} = 30\mathbb{Z} + 12\mathbb{Z} = 12\mathbb{Z} + 6\mathbb{Z} = 6\mathbb{Z},$$
$$\gcd(42, 30) = \gcd(30, 12) = \gcd(12, 6) = 6.$$

The following corollary is a rephrasing of the above theorem for practical uses.

Corollary 1.8 *Let d be the GCD of natural numbers a and b. There exist integers x and y that satisfy*

$$xa + yb = d.$$

Especially when a and b are coprime, that is $d = 1$, there exist integers x and y that satisfy

$$xa + yb = 1.$$

Proof Theorem 1.6 says $I = J$ if we set

$$I = \{xa + yb \mid x, y \in \mathbb{Z}\},$$
$$J = \{zd \mid z \in \mathbb{Z}\}.$$

Taking $z = 1$ in J, we know $d \in J$ and consequently $d \in I$, which means there are integers x and y that satisfy

$$xa + yb = d.$$

□

Example 1.9 We revisit Example 1.7. Let $a = 42$ and $b = 30$. The Euclidean algorithm goes like this:

$$(42\ 30)\begin{pmatrix} 0 & 1 \\ 1 & -1 \end{pmatrix} = (30\ 12),$$

$$(30\ 12)\begin{pmatrix} 0 & 1 \\ 1 & -2 \end{pmatrix} = (12\ 6),$$

$$(12\ 6)\begin{pmatrix} 0 & 1 \\ 1 & -2 \end{pmatrix} = (6\ 0).$$

Concatenating all these steps, we have

$$(42\ 30)\begin{pmatrix} 0 & 1 \\ 1 & -1 \end{pmatrix}\begin{pmatrix} 0 & 1 \\ 1 & -2 \end{pmatrix}\begin{pmatrix} 0 & 1 \\ 1 & -2 \end{pmatrix} = (6\ 0).$$

Calculating the product of the three square matrices, we have

$$(42\ 30)\begin{pmatrix} -2 & 5 \\ 3 & -7 \end{pmatrix} = (6\ 0),$$

which shows

$$-2 \times 42 + 3 \times 30 = 6.$$

Theorem 1.10 *If a prime p divides a product ab of two natural numbers a and b, then p divides a or b.*

Proof If p does not divide a, then $\gcd(a, p) = 1$. Corollary 1.8 claims that there exist integers x and y that satisfy

$$xp + ya = 1.$$

Multiplying both sides by p, we get

$$xpb + yab = b.$$

Since p divides ab, the prime p divides the LHS as a whole. Consequently, p divides the RHS, namely b.

Likewise, if p does not divide b, then p divides a. □

We have already shown that any natural numbers can be expressed as a product of primes. See Proposition 1.5. We will show that such a product is unique for each natural number.

Theorem 1.11 *Any natural number n can be expressed as a product of primes*

$$n = 2^{t_1} \times 3^{t_2} \times 5^{t_3} \times \cdots \times p_k^{t_k},$$

where $t_i \geq 0$ is an integer, and p_i is the i-th smallest prime. If n can be expressed as another product of primes

$$n = 2^{s_1} \times 3^{s_2} \times 5^{s_3} \times \cdots \times p_k^{s_k},$$

where $s_i \geq 0$ is an integer. Then, for all $1 \leq i \leq k$, we have $t_i = s_i$.

Quiz 1.2 The natural numbers k appearing in t_k and s_k are supposed to be equal. Does this cause any logical problems?

Proof Assume $t_1 \neq s_1$. We can further assume $t_1 > s_1$ without losing generality. If we divide n by 2^{s_1}, we get

$$2^{t_1 - s_1} \times 3^{t_2} \times 5^{t_3} \times \cdots \times p_k^{t_k} = 3^{s_2} \times 5^{s_3} \times \cdots \times p_k^{s_k}.$$

The natural number 2 divides the LHS and consequently the RHS. Take any prime from the RHS, say p, and set $n = pm$. We know that 2 does not divide $p > 2$ since p is a prime. Proposition 1.10 claims that 2 divides m. Take any prime factor from

m and repeat the same argument, then we come to the situation where 2 divides 1. This shows that we must have $t_1 = s_1$.

If we repeat this argument, after removing the power of 2, against

$$3^{t_2} \times 5^{t_3} \times \cdots \times p_k^{t_k} = 3^{s_2} \times 5^{s_3} \times \cdots \times p_k^{s_k},$$

then we get $t_2 = s_2$. If we further iterate this argument, we see that $t_i = s_i$ for all i. □

1.4 Polynomials

When we discuss a problem concerning polynomials of X, we must first set the domain of their coefficients. Frequently used domains in this book are the set of all integers $\mathbb{Z}$, rational numbers $\mathbb{Q}$, real numbers $\mathbb{R}$, and complex numbers $\mathbb{C}$.

Theorem 1.12 (Euclidean Division) *Let K be one of $\mathbb{Q}$, $\mathbb{R}$, or $\mathbb{C}$. Let $f(X)$ and $g(X)$ be polynomials of X with coefficients in K. There exist unique polynomials $q(X)$ and $r(X)$ with coefficients in K such that*

$$f(X) = q(X)g(X) + r(X),$$

where $\deg(r(X)) < \deg(g(X))$.

Remark 1.13 We say that the *quotient* of $f(X)$ divided by $g(X)$ is $q(X)$, and that the *remainder* is $r(X)$. If we take $K = \mathbb{Z}$ and want to get the same result, we must pose an additional condition that $g(X)$ is *monic*, namely the coefficient of its leading term is 1. The degree of the zero polynomial is often defined as -1 or $-\infty$. Either definition will do for our purpose.

Proof The *polynomial long division*, a familiar algorithm in high-school textbooks, tells us how to get the quotient and remainder. We only prove their uniqueness here. Suppose there is another pair of a quotient and remainder, say $q_1(X)$ and $r_1(X)$. We have

$$f(X) = q_1(X)g(X) + r_1(X).$$

Subtracting this from the former equation, we have

$$(q(X) - q_1(X))g(X) = r_1(X) - r(X).$$

The RHS has degree less than $\deg(g(X))$. We must have $q(X) - q_1(X) = 0$, otherwise the degree of the LHS would be greater than or equal to $\deg(g(X))$. Consequently, we have $r_1(X) = r(X)$. □

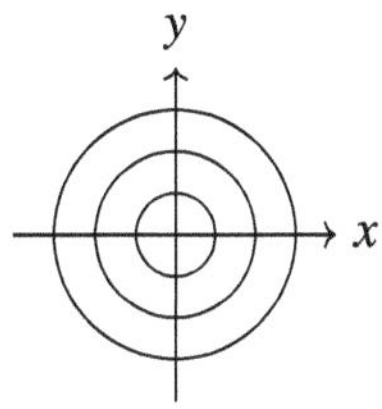

Fig. 2.6 Orbits of transformation group A

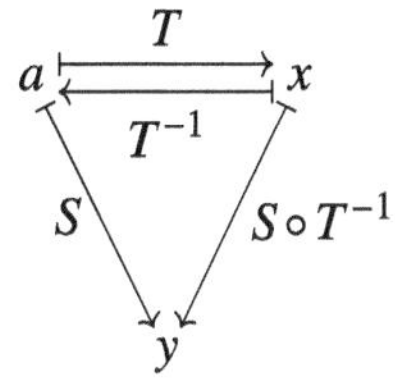

Fig. 2.7 Any orbit is transitive

Proof Suppose two orbits share a point, say a. The orbit that starts a travels over both orbits, which means the two orbits are united into a single orbit. □

Remark 2.11 Let a be a point in some orbit. Let x and y be any points that are moved from a by some transformations:

$$x = T(a), \quad y = S(a).$$

Then, we can find a transformation that sends x to y. From $x = T(a)$ we have

$$a = T^{-1}(x).$$

Substituting this in $y = S(a)$, we have

$$y = S(T^{-1}(x)) = (S \circ T^{-1})(x).$$

See Fig. 2.7. We can conclude that for any pair of points x and y on the same orbit there is always a transformation that sends x to y. Any point on an orbit can be referenced, at our disposal, as either the *starting point* or a *passing point* of the orbit.

Example 2.12 The following set B is a symmetry group acting on the xy-plane:

$$B = \left\{ \begin{pmatrix} a & 0 \\ 0 & a^{-1} \end{pmatrix} \middle| a > 0 \right\}.$$

Quiz 2.6 Show that B satisfies the four conditions for being a symmetry group.

The orbit of (x, y) is given by

$$\begin{pmatrix} a & 0 \\ 0 & a^{-1} \end{pmatrix} \begin{pmatrix} x \\ y \end{pmatrix} = \begin{pmatrix} ax \\ a^{-1}y \end{pmatrix},$$

letting a run in the range $a > 0$. Under the action of B, the product xy and the signature of x is kept constant. One branch of the hyperbola $xy = k$ with $x > 0$ is preserved. Another branch $xy = k$ with $x < 0$ is also preserved. If $k = 0$, it is a degenerate case to be treated with some care.

Definition 2.13 Let a transformation group G act on a set M. For any $m \in M$, we call the set

$$\{g(m) \mid g \in G\}$$

the G-orbit passing through m.

The positive part of the x-axis

$$I = \left\{ \begin{pmatrix} x \\ 0 \end{pmatrix} \middle| x > 0 \right\}$$

is a B-orbit. Let us see if this is true. Take any point on I, say $\begin{pmatrix} 1 \\ 0 \end{pmatrix}$, and apply any element of B:

$$\begin{pmatrix} a & 0 \\ 0 & a^{-1} \end{pmatrix} \begin{pmatrix} 1 \\ 0 \end{pmatrix} = \begin{pmatrix} a \\ 0 \end{pmatrix}.$$

If a runs through all the positive reals, the images run completely through I.

Quiz 2.7 Let k be a non-zero real number. Show that the following sets are B-orbits:

- the branch of the hyperbola $xy = k \quad (x > 0)$,
- the branch of the hyperbola $xy = k \quad (x < 0)$,
- the positive part of the x-axis,
- the negative part of the x-axis,
- the positive part of the y-axis,
- the negative part of the y-axis, and
- the singleton set consisting of the origin.

Collecting some B-orbits, we can construct a pattern in Fig. 2.8. Can we say that the symmetry group of this pattern is B? A careful look will make us find more symmetries in the exercises.

Fig. 2.8 *B* orbits

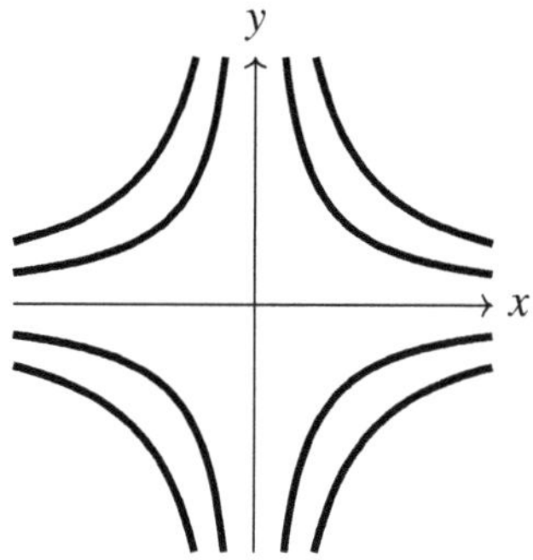

Exercises

[1] Find the symmetry group of each pattern below using the following transformations:

$$X_a : \begin{pmatrix} x \\ y \end{pmatrix} \mapsto \begin{pmatrix} x + a \\ y \end{pmatrix},$$

$$Y_a : \begin{pmatrix} x \\ y \end{pmatrix} \mapsto \begin{pmatrix} x \\ y + a \end{pmatrix},$$

$$R_a : \begin{pmatrix} x \\ y \end{pmatrix} \mapsto \begin{pmatrix} 2a - x \\ y \end{pmatrix},$$

$$S_a : \begin{pmatrix} x \\ y \end{pmatrix} \mapsto \begin{pmatrix} x \\ 2a - y \end{pmatrix}.$$

Y_a is the translation by a in the direction of the y-axis. S_a is the reflection with respect to the line $y = a$. Note that $X_0 = Y_0$, which we denote by I.

(1) Lines $\ell_1(y = 1)$ and $\ell_2(y = -1)$ with '☆'s on their points that have even integers as x-coordinates. See Fig. 2.9a.
(2) Lines $\ell_1(y = 1)$ and $\ell_2(y = -1)$ with '○'s on their points that have even integers as x-coordinates. See Fig. 2.9b.
(3) Lines $\ell_1(x = -1)$ and $\ell_2(x = +1)$. See Fig. 2.10a.
(4) Lines $\ell_1(x = 1)$ and $\ell_2(x = 2)$ with '$\wedge$'s on their lattice points. See Fig. 2.10b.
(5) A square with vertices at $(\pm 1, \pm 1)$. See Fig. 2.10c.

[2] Find a pattern preserved by each of the following transformation groups. Use lines $\ell_1(x = 1)$ and $\ell_2(x = -1)$ and place $\wedge$, □, or ○ on their points whose y-coordinates are even integers.

(1) $\{Y_a \mid a \in 2\mathbb{Z}\}$.

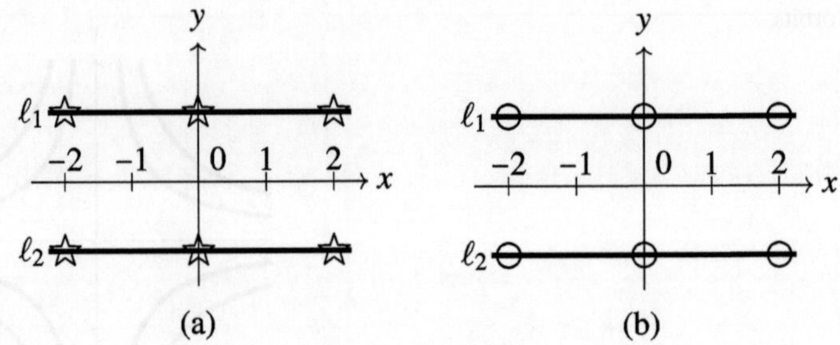

Fig. 2.9 Symbols on lattice points with even x-coordinates. (**a**) ☆. (**b**) ○

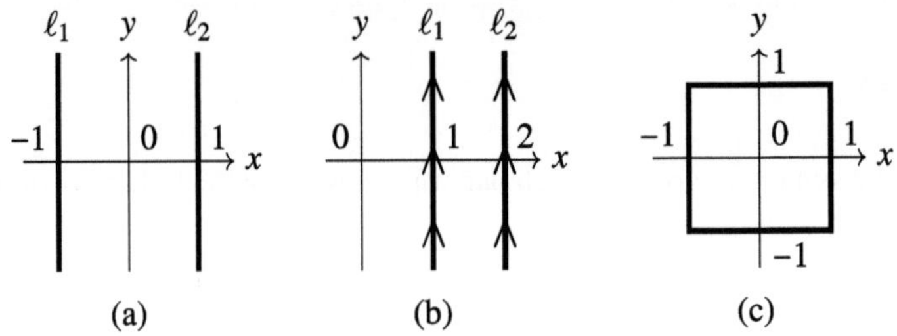

Fig. 2.10 Two parallel lines and a square. (**a**) Parallel lines. (**b**) With ∧. (**c**) A square

Fig. 2.11 An equilateral triangle in a circle

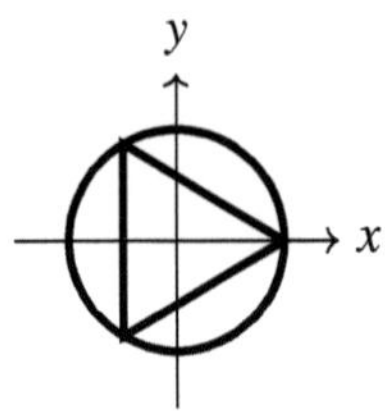

(2) $\{Y_a, S_b \mid a \in 2\mathbb{Z}, b \in \mathbb{Z}\}$.
(3) $\{Y_a, Y_a \circ R_0 \mid a \in 2\mathbb{Z}\}$.

[3] Let Z_θ be a transformation acting on the xy-plane:

$$Z_\theta : \begin{pmatrix} x \\ y \end{pmatrix} \mapsto \begin{pmatrix} \cos\theta & -\sin\theta \\ \sin\theta & \cos\theta \end{pmatrix} \begin{pmatrix} x \\ y \end{pmatrix}.$$

Consider the equilateral triangle and its circumcircle centered at the origin with radius 1. See Fig. 2.11. This pattern is preserved by $Z_0 = I$, $Z_{\frac{2\pi}{3}}$, and $Z_{\frac{4\pi}{3}}$.

(1) Find the matrix representations for the following transformations:

$$Z_{\frac{2\pi}{3}} \circ S_0, \qquad Z_{\frac{4\pi}{3}} \circ S_0,$$
$$S_0 \circ Z_{\frac{2\pi}{3}}, \qquad \text{and} \qquad S_0 \circ Z_{\frac{4\pi}{3}},$$

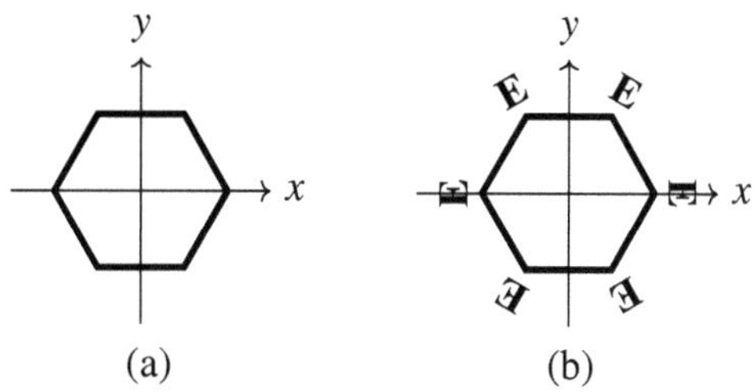

Fig. 2.12 Hexagons. (**a**) A hexagon only. (**b**) Decorated with **E**

where

$$S_0 : \begin{pmatrix} x \\ y \end{pmatrix} \mapsto \begin{pmatrix} 1 & 0 \\ 0 & -1 \end{pmatrix} \begin{pmatrix} x \\ y \end{pmatrix}.$$

(2) Find a symmetry group of the pattern in Fig. 2.11.

[4] Find the symmetry group of each pattern below. Use Z_θ and S_0.

(1) A hexagon. See Fig. 2.12a.
(2) A hexagon with 'E's at its vertices. See Fig. 2.12b.

[5] Consider the following transformation groups A and B:

$$A = \left\{ \begin{pmatrix} a & 0 \\ 0 & a^{-1} \end{pmatrix} \middle| a \in \mathbb{R}, a \neq 0 \right\},$$

$$B = \left\{ \begin{pmatrix} a & 0 \\ 0 & a^{-1} \end{pmatrix} \middle| a > 0 \right\}.$$

(1) Draw an A-orbit and B-orbit that pass through the point $(1, 1)$.
(2) Draw an A-orbit and B-orbit that pass through the point $(1, 0)$.

[6] Find the symmetry group for each pattern below. Use the linear transformations

$$\begin{pmatrix} a & 0 \\ 0 & b \end{pmatrix}, \quad ab = \pm 1.$$

(1) A pattern consisting of the hyperbolas $xy = \pm 1$ and $xy = \pm 2$. See Fig. 2.13a.
(2) A pattern consisting of three parabolas $xy = \pm 1$ and $xy = 2$. See Fig. 2.13b.
(3) A pattern consisting of the parabolas $xy = \pm 1$ and a branch of parabola

$$\begin{cases} xy = 2 \\ x > 0 \end{cases}, \quad \begin{cases} xy = -2 \\ x > 0 \end{cases}.$$

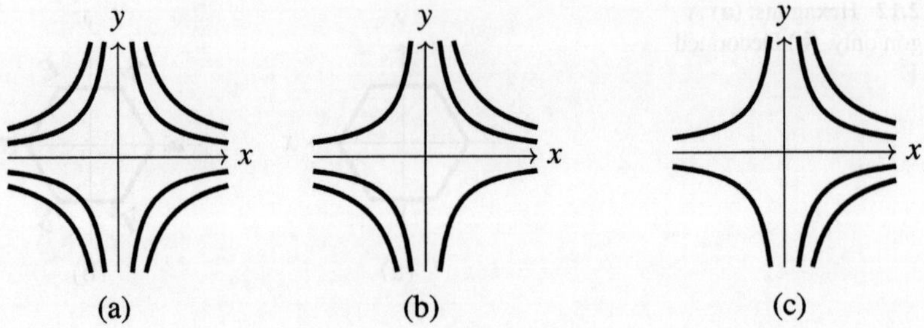

Fig. 2.13 An orbit may consist of several components. (**a**) $xy = \pm 1$, $xy = \pm 2$. (**b**) $xy = \pm 1$, $xy = 2$. (**c**) $xy = \pm 1$, $xy = \pm 2 (x > 0)$

See Fig. 2.13c.

[7] Find the orbit passing through the point $(1, 0)$ by each of the following transformation groups:

(1)

$$\left\{ \begin{pmatrix} \cos\theta & -\sin\theta \\ \sin\theta & \cos\theta \end{pmatrix} \middle| \theta = 0, \tfrac{2\pi}{3}, \tfrac{4\pi}{3} \right\}.$$

(2)

$$\left\{ \begin{pmatrix} \cos\theta & -\sin\theta \\ \sin\theta & \cos\theta \end{pmatrix} \middle| \theta = \tfrac{n\pi}{3}, n = 0, 1, \cdots, 5 \right\}.$$

(3)

$$\left\{ \begin{pmatrix} \cos\theta & -\sin\theta \\ \sin\theta & \cos\theta \end{pmatrix} \middle| 0 \leqq \theta < 2\pi \right\}.$$

Chapter 3
Groups and Orbits

So far, we have always considered transformation groups and the sets that are acted upon in pairs. In Sect. 3.1, we focus on algebraic structures of transformation groups leaving aside the sets acted upon. We introduce an abstract framework of group theory. In Sect. 3.2 , we introduce the concept of subgroups. In Sect. 3.3, we study the actions of groups on themselves, where the concept of coset decomposition or orbit decomposition plays the central role. In the final section, we study the dihedral group D_6 by applying all the techniques we have obtained so far.

3.1 Groups

Composition of transformations is associative. Are there any binary operations that do not observe the associative law? Yes, there are many. We can promptly present some examples.

Example 3.1

$$(3-1)-2 \neq 3-(1-2),$$
$$(3 \div 5) \div 2 \neq 3 \div (5 \div 2).$$

Subtraction and division are not associative. Advanced readers may think of examples in Lie algebras."

Definition 3.2 A set G equipped with a binary operation $\circ$ satisfying the following conditions is said to be a *group*.

(1) There is an *identity element*, or often called a *neutral element*, $e \in G$ having the property that $e \circ x = x \circ e = x$ for any $x \in G$.
(2) The binary operation $\circ$ is associative: for any $x, y, z \in G$, we have

S. Yukita, *First Lectures in Algebra*, Springer Asia Pacific Mathematics Series 10,
https://doi.org/10.1007/978-981-95-5457-7_3

$$(x \circ y) \circ z = x \circ (y \circ z).$$

(3) For any element $x \in G$, there is an *inverse* $y \in G$ such that

$$x \circ y = y \circ x = e.$$

Let us see the immediate consequences.

Proposition 3.3 *For any group G, there is only one neutral element e.*

Proof Let $e' \in G$ be another neutral element. It must satisfy condition (3) in Definition 3.2. Since e is a neutral element, we have

$$e \circ e' = e' \circ e = e'.$$

Since e' is also a neutral element, we have

$$e \circ e' = e' \circ e = e.$$

The above two equations show $e = e'$. □

Remark 3.4 The proof above is quite simple. But, we should note that it presents a powerful *proof pattern* when we want to show that something satisfying some conditions is unique. We first suppose there is another candidate. Then, we show that they turn out to be identical.

Proposition 3.5 *For any $x \in G$, there is only one inverse $y \in G$.*

Proof Let y_1, y_2 be two inverses of $x \in G$. We have

$$x \circ y_1 = e.$$

Composing y_2 from the left to both sides, we get

$$y_2 \circ x \circ y_1 = y_2.$$

Since $y_2 \circ x = e$, we have $y_1 = y_2$. □

From now on, we denote the unique inverse of x by x^{-1}. For brevity, we will write xy instead of $x \circ y$ whenever there can be no misunderstanding.

Let us see a series of new examples, permutation groups or symmetric groups.

Definition 3.6 Let $A = \{1, 2, 3\}$. Let us represent a bijection $\varphi : A \to A$ as

$$\begin{pmatrix} 1 & 2 & 3 \\ \varphi(1) & \varphi(2) & \varphi(3) \end{pmatrix}.$$

Such a bijection is often called a *permutation* of A. Let S_3 be the group of all the permutations of A under composition. We call it the *symmetric group of degree* 3:

$$S_3 = \left\{ \begin{pmatrix} 1 & 2 & 3 \\ 1 & 2 & 3 \end{pmatrix}, \begin{pmatrix} 1 & 2 & 3 \\ 2 & 3 & 1 \end{pmatrix}, \begin{pmatrix} 1 & 2 & 3 \\ 3 & 1 & 2 \end{pmatrix}, \begin{pmatrix} 1 & 2 & 3 \\ 2 & 1 & 3 \end{pmatrix}, \begin{pmatrix} 1 & 2 & 3 \\ 3 & 2 & 1 \end{pmatrix}, \begin{pmatrix} 1 & 2 & 3 \\ 1 & 3 & 2 \end{pmatrix} \right\}.$$

Composition is performed in the following way. For example, let

$$\sigma = \begin{pmatrix} 1 & 2 & 3 \\ 2 & 3 & 1 \end{pmatrix} \quad \text{and} \quad \tau = \begin{pmatrix} 1 & 2 & 3 \\ 2 & 1 & 3 \end{pmatrix}.$$

We calculate the composite point by point:

$$\begin{aligned} \sigma\tau(1) &= \sigma(\tau(1)) = \sigma(2) = 3, \\ \sigma\tau(2) &= \sigma(\tau(2)) = \sigma(1) = 2, \\ \sigma\tau(3) &= \sigma(\tau(3)) = \sigma(3) = 1. \end{aligned}$$

Gathering the above data, we express $\sigma\tau$ as

$$\sigma\tau = \begin{pmatrix} 1 & 2 & 3 \\ 3 & 2 & 1 \end{pmatrix}.$$

Note that we use the convention that τ applies first, then σ. Since permutations are mappings, their composition is associative.

Quiz 3.1 Find the neutral element of S_3. Find also the inverse of each permutation in S_3. This way, we can confirm that S_3 is a group.

Let us take a closer look at $\sigma = \begin{pmatrix} 1 & 2 & 3 \\ 2 & 3 & 1 \end{pmatrix}$. It is a permutation $1 \mapsto 2,\ 2 \mapsto 3,\ 3 \mapsto 1$. We can put it in a diagram

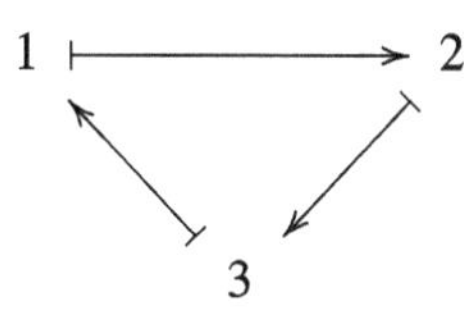

,

which makes it easier to see that σ is a cyclic permutation. Such a permutation is called simply a *cycle*. The cycle σ is often represented as $(1\ 2\ 3)$ for short.

Likewise, $\tau = \begin{pmatrix} 1 & 2 & 3 \\ 2 & 1 & 3 \end{pmatrix}$ can be put in a diagram

$$1 \rightleftarrows 2 \qquad 3\ .$$

Since 3 is a fixed point, the arrows in and out are omitted. It is a kind of cycle, but especially called a *transposition* and written (1 2) for short.

Let us represent the neutral element as e. Then, S_3 is

$$S_3 = \{e, (1\ 2\ 3), (1\ 3\ 2), (1\ 2), (1\ 3), (2\ 3)\}, \tag{3.1}$$

which eases the burden of writing.

Quiz 3.2 Show (3.1) holds.

Let us rewrite S_3 using $(1\ 2\ 3) = \sigma$, $(1\ 2) = \tau$, and their composites. One possibility is

$$S_3 = \{e, \sigma, \sigma^2, \tau, \sigma\tau, \sigma^2\tau\},$$

where we tried to place σ as far as possible to the left in each term.

Quiz 3.3 Show the following: $\sigma^2 = (1\ 3\ 2), \quad \tau\sigma = \sigma^2\tau, \quad \tau\sigma^2 = \sigma\tau.$

Remark 3.7 The latest rewriting of S_3 with σ and τ has a significant meaning. Using σ and τ, their composites cover all the elements of S_3. We say that σ and τ generate S_3. The set $\{\sigma, \tau\}$ is called a *system of generators*.

We will use the terms composition and multiplication interchangeably. The multiplication table of S_3 is given in Table 3.1.

In Table 3.1, look at the intersection of the row of τ and column of σ. At this entry we must have $\tau\sigma$. There are many representations for this element. If we adopt the convention that σ must be moved to the left, the preferred representative is $\sigma^2\tau$.

The rightmost column has the title "order". This is the least natural number n such that $x^n = e$ for x in the title column. The order of e is defined as 1.

Table 3.1 Multiplication table of S_3

	R							
L	e	σ	σ^2	τ	$\sigma\tau$	$\sigma^2\tau$	Inverse	Order
e	e	σ	σ^2	τ	$\sigma\tau$	$\sigma^2\tau$	e	1
σ	σ	σ^2	e	$\sigma\tau$	$\sigma^2\tau$	τ	σ^2	3
σ^2	σ^2	e	σ	$\sigma^2\tau$	τ	$\sigma\tau$	σ	3
τ	τ	$\sigma^2\tau$	$\sigma\tau$	e	σ^2	σ	τ	2
$\sigma\tau$	$\sigma\tau$	τ	$\sigma^2\tau$	σ	e	σ^2	$\sigma\tau$	2
$\sigma^2\tau$	$\sigma^2\tau$	$\sigma\tau$	τ	σ^2	σ	e	$\sigma^2\tau$	2

3.2 Subgroups

Each of the following subsets of S_3 satisfies all the requirements for being a group.

$$\{e\}, \quad \{e, \tau\}, \quad \{e, \sigma\tau\}, \quad \{e, \sigma^2\tau\}, \quad \{e, \sigma, \sigma^2\}, \quad S_3.$$

These are called a *subgroup* of S_3.

Quiz 3.4 Show that each of the above subsets of S_3 qualifies for being a group by making the multiplication table.

Let us see a formal definition.

Definition 3.8 Let G be a group, and let H be a subset of G. The subset H is called a *subgroup* of G if it satisfies all the following conditions:

(1) $e \in H$,
(2) $xy \in H$ for any $x, y \in H$, and
(3) $x^{-1} \in H$ for any $x \in H$.

Remark 3.9 Both the subsets $\{e\}$ and G are considered subgroups of G. They are called *trivial subgroups*. Condition (1) can be removed since it can be deduced from (2) and (3).

Let H_1, H_2 be subgroups of G. Is the intersection $H_1 \cap H_2$ a subgroup of G? Is the union $H_1 \cup H_2$ a subgroup of G?

Theorem 3.10 *If H_1 and H_2 are subgroups of G, then their intersection $H_1 \cap H_2$ is also a subgroup of G.*

Proof For any $x, y \in H_1 \cap H_2$, we have

$$xy \in H_1 \quad \text{and} \quad xy \in H_2,$$

since H_1 and H_2 are subgroups. Thus, we have

$$xy \in H_1 \cap H_2.$$

The inverse satisfies

$$x^{-1} \in H_1 \quad \text{and} \quad x^{-1} \in H_2.$$

Therefore,

$$x^{-1} \in H_1 \cap H_2.$$

□

Quiz 3.5 Let H_1 and H_2 be subgroups of G. Their union $H_1 \cup H_2$ is not a subgroup of G in general. Set $H_1 = \{e, \tau\}$ and $H_2 = \{e, \sigma, \sigma^2\}$. See if $H_1 \cup H_2$ is a subgroup of S_3.

Let us take a closer look at the multiplication Table 3.1 of S_3. We find that on each row, excluding the title column, all the elements of S_3 appear exactly once. For example, take the row of σ in the title column. The row contains the results of multiplication by σ from the left with all the elements of S_3. We denote such a left multiplication by $\sigma\bullet$.

$$\sigma\bullet : \begin{array}{cccccc} e & \sigma & \sigma^2 & \tau & \sigma\tau & \sigma^2\tau \\ \downarrow & \downarrow & \downarrow & \downarrow & \downarrow & \downarrow \\ \sigma & \sigma^2 & e & \sigma\tau & \sigma^2\tau & \tau \end{array}$$

The mapping $\sigma\bullet : S_3 \to S_3$ is a bijection, thus it is a permutation of the set S_3. Proposition 3.11 will show that multiplications from the left by other elements of S_3 have this property like σ. Using this fact, we see that the set

$$S_3\bullet = \{e\bullet, \quad \sigma\bullet, \quad \sigma^2\bullet, \quad \tau\bullet, \quad (\sigma\tau)\bullet, \quad (\sigma^2\tau)\bullet\}$$

forms a transformation group acting on the set S_3. The group S_3 has been defined as a transformation group acting on the set $\{1, 2, 3\}$. This time S_3 plays the role of the set (space) acted upon by $S_3\bullet$.

Proposition 3.11 *Let G be a group. For any $g \in G$ the left multiplication $g\bullet : G \to G$ is a bijection, namely a permutation of G.*

Proof We first show that $g\bullet$ is injective. Suppose $g \bullet (x_1) = g \bullet (x_2)$ for $x_1, x_2 \in G$. By definition we have $gx_1 = gx_2$. If we multiply both sides by g^{-1} from the left, then we have $x_1 = x_2$. Thus, $g\bullet : G \to G$ is injective.

Next, we show that $g\bullet$ is surjective. For any $y \in G$, set $x = g^{-1}y$. Then, we have $g \bullet (x) = gx = gg^{-1}y = y$. Thus, $g\bullet : G \to G$ is surjective. □

Proposition 3.12 *Let G be a group.*

(i) Let $e \in G$ be the neutral element of G. Then, $e\bullet : G \to G$ is the identity transformation.
(ii) For any $g_1, g_2 \in G$, we have $(g_2\bullet) \circ (g_1\bullet) = (g_2g_1)\bullet$.
(iii) For any $g \in G$, we have $(g\bullet)^{-1} = (g^{-1})\bullet$.

Proof (i) is trivial.

We show (ii) by pointwise calculation. Let $x \in G$. We have the following:

$$((g_2\bullet) \circ (g_1\bullet))(x) = (g_2\bullet)(g_1 \bullet x) = g_2 \bullet (g_1x) = g_2g_1x,$$

$$(g_2g_1) \bullet (x) = g_2g_1x.$$

This shows

$$((g_2\bullet) \circ (g_1\bullet))(x) = (g_2 g_1) \bullet (x).$$

Since $x \in G$ is arbitrary, we have

$$(g_2\bullet) \circ (g_1\bullet) = (g_2 g_1) \bullet .$$

Finally, we show (iii). By (ii) we have

$$(g\bullet) \circ (g^{-1}\bullet) = (g^{-1}\bullet) \circ (g\bullet) = e \bullet .$$

By (i) we know that $e\bullet$ is the identity, namely the neutral element. Therefore, $(g\bullet)^{-1} = g^{-1}\bullet$. □

Remark 3.13 Let us denote the transformation group $\{g \bullet \mid g \in G\}$ by $G\bullet$. An obvious correspondence $g \mapsto g\bullet$ gives a bijection from G to the transformation group $G\bullet$. The above proposition claims that this bijection preserves the binary operation. Later, we will introduce the concept of *isomorphisms*. We will say that G and $G\bullet$ are isomorphic. So, we often identify G with $G\bullet$ and say that G acts on G with left multiplication.

The following is a corollary of Proposition 3.12.

Proposition 3.14 *Let G be a group. There is only one $G\bullet$-orbit, namely the whole set G.*

Proof Let $g \in G$. We show that the $G\bullet$-orbit that starts g is the whole G. Given any $h \in G$, the transformation $(hg^{-1})\bullet$ sends g to h as follows:

$$(hg^{-1}) \bullet (g) = hg^{-1}g = h.$$

□

3.3 Orbit or Coset Decomposition

Let G be a group and H be a subgroup of G. By left multiplication, H can be seen as a transformation group acting on the set G. To be precise, if we set

$$H\bullet = \{h \bullet \mid h \in H\},$$

we get a transformation group. We study various $H\bullet$-orbits.

Definition 3.15 Let G be a group. Let H be a subgroup of G. An $H\bullet$ orbit is called a *left* H-orbit. We denote a left H-orbit that passes $x \in G$ by Hx.

Table 3.2 Left A-orbits

Passes e	
$A\bullet$	e
e	e
σ	σ
σ^2	σ^2

Passes τ	
$A\bullet$	τ
e	τ
σ	$\sigma\tau$
σ^2	$\sigma^2\tau$

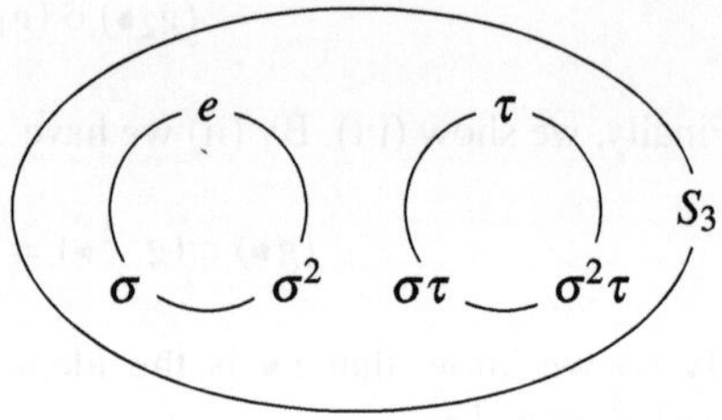

Fig. 3.1 Left A-orbits

Table 3.3 Left B-orbits

Passes e	
$B\bullet$	e
e	e
τ	τ

Passes σ	
$B\bullet$	σ
e	σ
τ	$\sigma^2\tau$

Passes σ^2	
$B\bullet$	σ^2
e	σ^2
τ	$\sigma\tau$

Remark 3.16 In most textbooks, a left H-orbit Hx is often called a *right coset* with *representative* x. The representative x appears on the right in Hx. So, it has the name. However, some textbooks adopt the opposite convention, namely *left coset.*

We study S_3 and its subgroup $A = \{e, \sigma, \sigma^2\}$. The subgroup A acts on the set S_3 by the left multiplication. There are two left A-orbits: one passing e and the other τ.

$$Ae = \{e, \sigma, \sigma^2\},$$
$$A\tau = \{\tau, \sigma\tau, \sigma^2\tau\}.$$

See Table 3.2.
Figure 3.1 shows that S_3 is decomposed into two orbits.
Next, we consider a subgroup $B = \{e, \tau\}$. There are three left B-orbits: one passing e, one passing σ, and the other passing σ^2.

$$Be = \{e, \tau\},$$
$$B\sigma = \{\sigma, \sigma^2\tau\},$$
$$B\sigma^2 = \{\sigma^2, \sigma\tau\}.$$

See Table 3.3. Figure 3.2 shows that S_3 is decomposed into three orbits.

Fig. 3.2 Left B-orbits

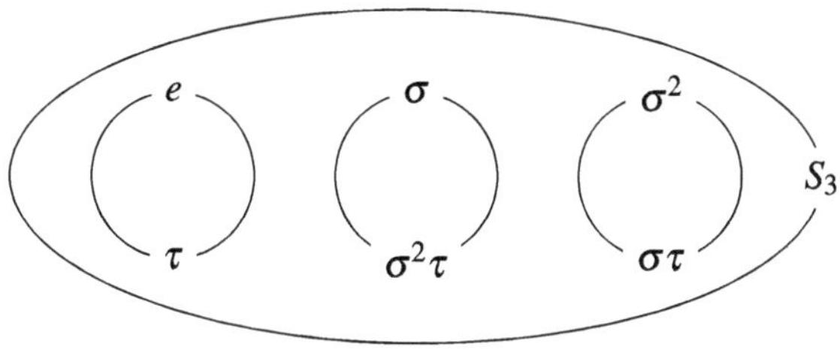

In general, if H is a subgroup of G, the left H-orbit starting at x is denoted by Hx. Namely, we have

$$Hx = \{hx \mid h \in H\}.$$

Take any other element on the orbit Hx, say h_0x $(h_0 \in H)$. The left H-orbit starting at h_0x is calculated as

$$H(h_0x) = \{hh_0x \mid h \in H\} = \{hx \mid h \in H\},$$

where we used the fact that hh_0 runs through exactly the whole H when h runs through H. This observation reassures us of the following:

(i) We can say "a left H-orbit starting at x" and "a left H-orbit passing x" interchangeably.
(ii) Representative of a left H-orbit can be chosen arbitrarily. Let $y \in Hx$. Then, we have $Hx = Hy$.

We denote by $H\backslash G$ the set of all the left H-orbits. We can write

$$A\backslash S_3 = \{A, A\tau\} \quad \text{and} \quad B\backslash S_3 = \{B, B\sigma, B\sigma^2\}.$$

We give some remarks on terminology and notation. Let X be a set. We denote by $|X|$ the *size* of the set X. We use the word "size" instead of "number" or "cardinality" just for ease of reading. The size of a group is usually called the *order*, which has overloaded meanings. The author wants to avoid such a situation for an introductory book.

Observations so far tempt us to conjecture the following.

Theorem 3.17 (Lagrange) *Let H be a subgroup of G.*

(i) The set G is the disjoint union of all the left H-orbits.
(ii) The left H-orbit passing e is the set H.
(iii) The size (number, cardinality) of every left H-orbit is equal to $|H|$.
(iv) If $|G| < \infty$, then $|H|$ divides $|G|$. In other words, $|H|$ is a divisor of $|G|$.

Proof We show (i). For any $x \in G$, there is a left H-orbit Hx. Therefore, the union of all left H-orbits is G. By Proposition 2.10 we know that left orbits are mutually disjoint.

(ii) is trivial.

We show (iii). Take any pair of orbits Hx_1 and Hx_2. We can define a mapping $f : Hx_1 \to Hx_2$ by

$$f : hx_1 \mapsto hx_2$$

for all $h \in H$. This mapping is just the multiplication from the right by $x_1^{-1}x_2$. The inverse is the multiplication from the right by $x_2^{-1}x_1$. So, f is a bijection, which means $|Hx_1| = |Hx_2|$.

We show (iv). From (iii), we know that all left H-orbits have the same size as H. Therefore, we have

$$|G| = |H| \times |H \backslash G|.$$

□

Theorem 3.18 *Let G be a group. If $|G| < \infty$, then for any $x \in G$ its order is a divisor of $|G|$.*

Proof Assume the order of x is infinity. Then,

$$x, x^2, x^3, \ldots$$

are all mutually different. Since this kind of situation is not allowed if $|G| < \infty$, we can claim that the order of x is finite.

Let the natural number n be the order of x. Then, we have a subgroup

$$\{e, x, \ldots, x^{n-1}\},$$

whose size is n. From Theorem 3.17, we see that n is a divisor of $|G|$. □

From now on, we rotate our viewpoint by 90°. Let us take a careful look at each column in Table 3.1. We notice that every element of S_3 appears exactly once in the column, which means that the multiplication by a fixed element from the right is bijective.

Let G be a group. For any $g \in G$, we define the multiplication by g from the right $\bullet g : G \to G$ by

$$\bullet g : x \mapsto xg.$$

Proposition 3.19 *Let G be a group. For any g, the right multiplication $\bullet g : G \to G$ is a bijection, namely a permutation of G.*

Proof We first show that $\bullet g$ is injective. Suppose $\bullet g(x_1) = \bullet g(x_2)$ for $x_1, x_2 \in G$. By definition we have $x_1 g = x_2 g$. If we multiply both sides by g^{-1} from the right, then we have $x_1 = x_2$. Thus, $\bullet g : G \to G$ is injective.

Next, we show that $\bullet g$ is surjective. For any $y \in G$, set $x = yg^{-1}$. Then, we have $\bullet g(x) = xg = yg^{-1}g = y$. Thus, $\bullet g : G \to G$ is surjective. □

Proposition 3.20 *Let G be a group.*

(i) Let $e \in G$ be the neutral element of G. Then, $\bullet e : G \to G$ is the identity transformation.
(ii) For any $g_1, g_2 \in G$, we have $(\bullet g_2) \circ (\bullet g_1) = \bullet(g_1 g_2)$. Note that the orders of g_1 and g_2 are different on the LHS and RHS.
(iii) For any $g \in G$, we have $(\bullet g)^{-1} = \bullet(g^{-1})$.

Proof (i) is trivial.

We show (ii) by pointwise calculation. Let $x \in G$. We have the following:

$$((\bullet g_2) \circ (\bullet g_1))(x) = (\bullet g_2)((\bullet g_1)(x)) = (\bullet g_2)(x g_1) = x g_1 g_2,$$
$$(\bullet(g_1 g_2))(x) = x g_1 g_2.$$

This shows

$$((\bullet g_2) \circ (\bullet g_1))(x) = (\bullet(g_1 g_2))(x).$$

Since $x \in G$ is arbitrary, we have

$$(\bullet g_2) \circ (\bullet g_1) = \bullet(g_1 g_2).$$

Finally, we show (iii). By (ii) we have

$$(\bullet g) \circ (\bullet g^{-1}) = \bullet(g^{-1} g) = \bullet e.$$

By (i) we know that $\bullet e$ is the identity, namely the neutral element. Therefore, $(\bullet g)^{-1} = \bullet g^{-1}$. □

Remark 3.21 We denote the transformation group $\{\bullet g \mid g \in G\}$ by $\bullet G$. An obvious correspondence $g \mapsto \bullet g$ gives a bijection from G to the transformation group $\bullet G$. The above proposition claims that this bijection nearly preserves the binary operation. The only setback is that order of elements is reversed. We say that G and $\bullet G$ are *anti-isomorphic*. Although this subtlety exists, we can deal with orbits of right multiplication in the same way as left multiplication.

Definition 3.22 Let G be a group. Let H be a subgroup of G. An $\bullet H$ orbit is called a *right* H-orbit. We denote a right H-orbit that passes $x \in G$ by xH.

Table 3.4 Right A-orbits

Starts at e

	$\bullet A$		
	e	σ	σ^2
e	e	σ	σ^2

Starts at τ

	$\bullet A$		
	e	σ	σ^2
τ	τ	$\sigma^2\tau$	$\sigma\tau$

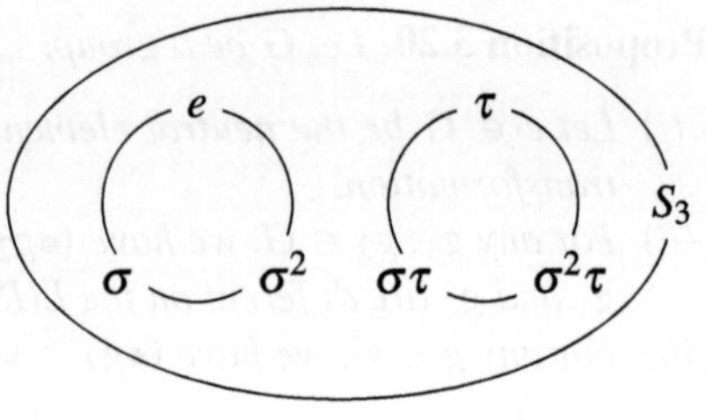

Fig. 3.3 Right A-orbits

Table 3.5 Right B-orbits

Starts at e

	$\bullet B$	
	e	τ
e	e	τ

Starts at σ

	$\bullet B$	
	e	τ
σ	σ	$\sigma\tau$

Starts at σ^2

	$\bullet B$	
	e	τ
σ^2	σ^2	$\sigma^2\tau$

Remark 3.23 In most textbooks, a right H-orbit xH is called a *left coset* with *representative* x. The representative x appears one the left in xH. So, it has the name. However, some textbooks adopt the opposite convention, namely *right coset*.

We study S_3 and the following subgroups:

$$A = \{e, \sigma, \sigma^2\},$$
$$B = \{e, \tau\}.$$

Let us calculate right A-orbits and right B-orbits.

There are two right A-orbits, namely $\{e, \sigma, \sigma^2\}$ and $\{\tau, \sigma\tau, \sigma^2\tau\}$. See Table 3.4. We use the notation below:

$$\{e, \sigma, \sigma^2\} = eA = A, \quad \{\tau, \sigma^2\tau, \sigma\tau\} = \tau A.$$

See Fig. 3.3 for the decomposition by right A-orbits.

There are three right B-orbits, namely $\{e, \tau\}$, $\{\sigma, \sigma\tau\}$, and $\{\sigma^2, \sigma^2\tau\}$. See Table 3.5. We use the notation below:

$$\{e, \tau\} = eB = B, \quad \{\sigma, \sigma\tau\} = \sigma B, \quad \{\sigma^2, \sigma^2\tau\} = \sigma^2 B.$$

See Fig. 3.4 for the decomposition by right B-orbits.

In general, if H is a subgroup of G, the right H-orbit starting at x is denoted by xH. Namely, we have

Fig. 3.4 Right B-orbits

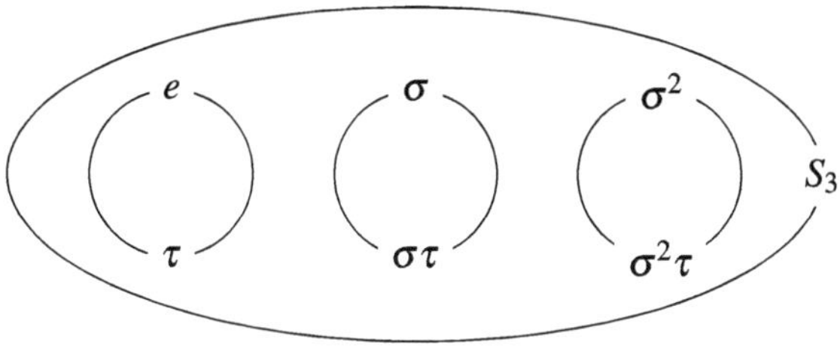

$$xH = \{xh \mid h \in H\}.$$

Take any other element on the orbit xH, say xh_0 ($h_0 \in H$). The right H-orbit starting at xh_0 is calculated as

$$(xh_0)H = \{xh_0h \mid h \in H\} = \{xh \mid h \in H\},$$

where we used the fact that h_0h runs through exactly the whole H when h runs through H. This observation reassures us of the following:

(i) We can say "a right H-orbit starting at x" and "a right H-orbit passing x" interchangeably.
(ii) Representative of a right H-orbit can be chosen arbitrarily. Let $y \in xH$. Then, we have $xH = yH$.

We denote by G/H the set of all the right H-orbits. We can write

$$S_3/A = \{A, \tau A\} \quad \text{and} \quad S_3/B = \{B, \sigma B, \sigma^2 B\}.$$

We have a theorem on right orbits like that on left orbits, Theorem 3.17.

Theorem 3.24 (Lagrange) *Let H be a subgroup of G.*

(i) The set G is the disjoint union of all the right H-orbits.
(ii) The right H-orbit passing e is the set H.
(iii) The size of every right H-orbit is equal to $|H|$.
(iv) If $|G| < \infty$, then $|H|$ divides $|G|$. (We have already known this.)

Proof We show (i). For any $x \in G$, there is a right H-orbit xH. Therefore, the union of all right H-orbits is G. By Proposition 2.10 we know that right orbits are mutually disjoint.

(ii) is trivial.

We show (iii). Take any pair of orbits x_1H and x_2H. We can define a mapping $f : x_1H \to x_2H$ by

$$f : x_1h \mapsto x_2h$$

for all $h \in H$. This mapping is just the multiplication from the left by $x_2 x_1^{-1}$. The inverse is the multiplication from the left by $x_1 x_2^{-1}$. So, f is a bijection, which means $|x_1 H| = |x_2 H|$.

We show (iv). From (iii), we know that all right H-orbits have the same size as H. Therefore, we have

$$|G| = |H| \times |G/H|.$$

□

Now, we return to the concrete example of S_3 and its subgroup $A = \{e, \sigma, \sigma^2\}$. By calculation, we have

$$Ae = eA \quad \text{and} \quad A\tau = \tau A,$$

which shows that the left A-orbit and right A-orbit that start at the same point coincide. Therefore, the right orbit decomposition and the left are identical. We express this fact as

$$A\backslash S_3 = S_3/A.$$

Quiz 3.6 Confirm that $A\backslash S_3 = S_3/A$.

Let us see if the same is true for the subgroup $B = \{e, \tau\}$. Direct calculation shows

$$Be = eB, \quad B\sigma \neq \sigma B, \quad \text{and} \quad B\sigma^2 \neq \sigma^2 B.$$

The left and right orbits that start at the same point may differ. Therefore the left and right orbit decompositions differ. We express this fact as

$$B\backslash S_3 \neq S_3/B.$$

Quiz 3.7 Confirm that $B\backslash S_3 \neq S_3/B$.

Definition 3.25 Let H be a subgroup of G. The subgroup H is called a *normal subgroup* if the left and right H-orbit decompositions coincide. Namely,

$$H\backslash G = G/H.$$

Remark 3.26 Trivial subgroups $\{e\}$ and G are normal subgroups of G. The condition in Definition 3.25 can be rephrased as follows. For any $x \in G$, the left and right H-orbits coincide. Namely, $Hx = xH$. The role of normal subgroups will be fully discussed in Sect. 4.2.

The subgroup $A = \{e, \sigma, \sigma^2\}$ is a normal subgroup of S_3, but the subgroup $B = \{e, \tau\}$ is not.

Quiz 3.8 Consider the subgroup $B_1 = \{e, \sigma\tau\}$ and $B_2 = \{e, \sigma^2\tau\}$ of S_3. Show that neither of them is a normal subgroup by explicitly writing down left and right orbit decompositions.

Example 3.27 Let $\mathbb{Z} = \{0, \pm 1, \pm 2, \pm 3, \ldots\}$ be the set of all integers. The set $\mathbb{Z}$ is considered a group with addition as a binary operation. We write $x + y$ instead of xy or $x \circ y$. In most examples we have seen so far, $xy \neq yx$ in general. For $\mathbb{Z}$, we always have $x + y = y + x$. Such a group is called a *commutative group* or *abelian group*. The group $\mathbb{Z}$ has many subgroups. For example,

$$3\mathbb{Z} = \{3n \mid n \in \mathbb{Z}\},$$

the set of all multiples of 3, is a subgroup.

For abelian groups, we cannot distinguish the left and right orbits. Therefore, all subgroups are normal subgroups. So, we just say orbits and cosets. There is no ambiguity when we write

$$\mathbb{Z}/3\mathbb{Z} = \{3\mathbb{Z}, 3\mathbb{Z} + 1, 3\mathbb{Z} + 2\},$$

where we denote by

$$3\mathbb{Z} + 1 = \{3n + 1 \mid n \in \mathbb{Z}\}$$

the coset with representative 1, or $3\mathbb{Z}$-orbit starting at 1. Since the distinction between "left" and "right" is meaningless in abelian groups, we can use

$$\mathbb{Z}/3\mathbb{Z} \quad \text{and} \quad 3\mathbb{Z}\backslash\mathbb{Z}$$

interchangeably.

Example 3.28 A multiplicative representation of the group $\mathbb{Z}$ may be given by

$$C_\infty = \{\ldots, a^{-2}, a^{-1}, e, a, a^2, \ldots\}.$$

The subgroup corresponding to $3\mathbb{Z}$ of $\mathbb{Z}$ is

$$A = \{a^{3n} \mid n \in \mathbb{Z}\}.$$

Then, we have

$$C_\infty/A = \{C_\infty, aC_\infty, a^2C_\infty\}.$$

Definition 3.29 An element g of a group G is called a *generator* if

$$G = \{e, g^{\pm 1}, g^{\pm 2}, g^{\pm 3}, \cdots\},$$

where the enumeration may terminate in finite iterations. A group G is called a *cyclic group* if it has a generator. If G has a generator g of finite order, say n, we have

$$G = \{e, g, g^2, \cdots, g^{n-1}\},$$

where $g^{n-1} = g^{-1}, g^{n-2} = g^{-2}, \cdots, g = g^{-(n-1)}$. In this case G is called a *finite cyclic group*. If the enumeration does not terminate, then G is called a *infinite cyclic group*.

Example 3.30 Examples of cyclic groups are given below.

(1) $\mathbb{Z}$ is a cyclic group with an obvious generator 1. Note that -1 is also a generator.
(2) C_∞ is a cyclic group with an obvious generator a. We can also take a^{-1} as a generator. $\mathbb{Z}$ and C_∞ are isomorphic (we have not defined this term yet).
(3) $C_6 = \{e, a, a^2, a^3, a^4, a^5\}$ is a group under the rules $a^6 = e$ and $a^m a^n = a^{m+n}$. Then, C_6 is a cyclic group with an obvious generator a. We can also take $a^{-1} = a^5$ as a generator.

Quiz 3.9 Show that the element a^5 is a generator of the group $C_6 = \{e, a, a^2, a^3, a^4, a^5\}$.

Proposition 3.31 *Let G be a finite group with $n = |G|$. If there is an element $x \in G$ of order n, then x is a generator of G.*

Proof The elements $e, x, x^2, \cdots, x^{n-1}$ are all different. Otherwise, there is a pair $x^k = x^\ell$ $(0 \leq k < \ell < n)$, and we have $x^{\ell-k} = e$ $(1 \leq \ell - k < n)$, which contradicts the assumption that x has order n.

Thus, we know

$$\{e, x, x^2, \cdots, x^{n-1}\} \subset G.$$

Both sets has size 6. Therefore, we have

$$\{e, x, x^2, \cdots, x^{n-1}\} = G.$$

□

Quiz 3.10 The group S_3 is not a cyclic group since it has no generators. Show this by calculating the order of each element.

Definition 3.32 Let G be a group. A subset $A \subset G$ is said to *generate* G and called a *system of generators* if repeating multiplication among elements of A and their inverses produces all the elements of G.

Example 3.33 The sets $\{\sigma, \tau\}$ and $\{(1\ 2), (1\ 3)\}$ are examples of a system of generators of S_3.

Quiz 3.11 Show that $\{(1\ 2), (1\ 3)\}$ is a system of generators of S_3.

3.4 Dihedral Group D_6

We will study the structures of D_6 using only the basic knowledge introduced so far.

The dihedral group D_6 is given by

$$D_6 = \{e, a, \cdots, a^5, b, ab, a^2b, \cdots, a^5b\},$$

with the following rules:

$$ba = a^5b, \; ba^2 = a^4b, \; ba^3 = a^3b,$$
$$ba^4 = a^2b, \; ba^5 = ab, \; a^6 = b^2 = e.$$

Table 3.6 shows closedness and the existence of the identity and inverses. What remains to be proven is associativity.
A brute-force way is to examine a total of $12^3 = 1728$ equalities

$$(xy)z = x(yz)$$

for all triplets $(x, y, z) \in D_6 \times D_6 \times D_6$. The reader will agree that this task is for a computer.

The group D_n $(n \geq 3)$ is named a dihedral group after its geometrical origin. It arises as the group of symmetries of the regular n-gon. The element a represents the rotation around the center by $\frac{2\pi}{n}$. The element b represents the reflection with respect to an antipodal diagonal or midpoint connection line. See Fig. 3.5.
Based on this background, associativity is automatic. Revisit Proposition 2.6.

The reader who finds the following quiz and subsequent remark puzzling can safely skip to the ¶sign below.

Table 3.6 Table for D_6

e	a	a^2	a^3	a^4	a^5	b	ab	a^2b	a^3b	a^4b	a^5b	inv.	o.	
e	e	a	a^2	a^3	a^4	a^5	b	ab	a^2b	a^3b	a^4b	a^5b	e	1
a	a	a^2	a^3	a^4	a^5	e	ab	a^2b	a^3b	a^4b	a^5b	b	a^5	6
a^2	a^2	a^3	a^4	a^5	e	a	a^2b	a^3b	a^4b	a^5b	b	ab	a^4	3
a^3	a^3	a^4	a^5	e	a	a^2	a^3b	a^4b	a^5b	b	ab	a^2b	a^3	2
a^4	a^4	a^5	e	a	a^2	a^3	a^4b	a^5b	b	ab	a^2b	a^3b	a^2	3
a^5	a^5	e	a	a^2	a^3	a^4	a^5b	b	ab	a^2b	a^3b	a^4b	a	6
b	b	a^5b	a^4b	a^3b	a^2b	ab	e	a^5	a^4	a^3	a^2	a	b	2
ab	ab	b	a^5b	a^4b	a^3b	a^2b	a	e	a^5	a^4	a^3	a^2	ab	2
a^2b	a^2b	ab	b	a^5b	a^4b	a^3b	a^2	a	e	a^5	a^4	a^3	a^2b	2
a^3b	a^3b	a^2b	ab	b	a^5b	a^4b	a^3	a^2	a	e	a^5	a^4	a^3b	2
a^4b	a^4b	a^3b	a^2b	ab	b	a^5b	a^4	a^3	a^2	a	e	a^5	a^4b	2
a^5b	a^5b	a^4b	a^3b	a^2b	ab	b	a^5	a^4	a^3	a^2	a	e	a^5b	2

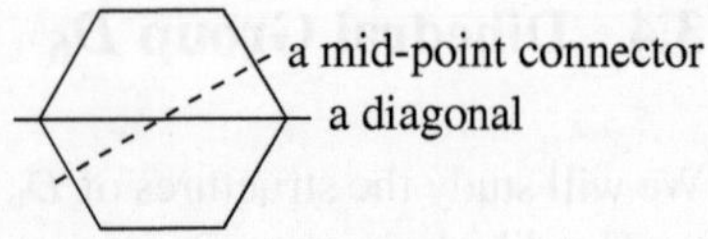

Fig. 3.5 Antipodal reflections

Quiz 3.12 Fill out the condensed multiplication table of D_6 below. Show that D_6 is closed under multiplication and inversion. Show the existence of the neutral element.

	a^j	$a^j b$	inverse	order
a^i				$\varphi(i)$
$a^i b$				2

Here, we assume $0 \leq i, j < 5$. Find the value of $\varphi(i)$ for each $i = 0, 1, \cdots, 5$.

Remark 3.34 The reader who accomplished the above quiz will find that the multiplication rules are unified into a single formula

$$a^i b^u \cdot a^j b^v = a^{i+(-1)^u j} b^{u+v}, \tag{3.2}$$

concise but a little bit tedious for our eyes. Here, the exponents of a are calculated modulo 6. The exponents of b are modulo 2. Examples are given below:

$$a^4 b \cdot a = a^4 b \cdot ab^0 = a^{4+(-1)^1} b^{1+0} = a^3 b,$$

$$ab \cdot a^3 b = a^1 b^1 \cdot a^3 b^1 = a^{1+(-1)^1 \times 3} b^{1+1} = a^{-2} b^2 = a^4.$$

To confirm the associative law from scratch, we must do the following. Take any triplet of elements of D_6:

$$a^i b^u, \quad a^j b^v, \quad \text{and} \quad a^k b^w.$$

Then, show the equation

$$(a^i b^u \cdot a^j b^v) \cdot a^k b^w = a^i b^u \cdot (a^j b^v \cdot a^k b^w),$$

by calculation. Using Eq. (3.2), we can proceed step by step as follows:

$$\begin{aligned}
\text{LHS} &= a^{i+(-1)^u j} b^{u+v} \cdot a^k b^w \\
&= a^{i+(-1)^u j+(-1)^{u+v} k} b^{u+v+w}, \\
\text{RHS} &= a^i b^u \cdot a^{j+(-1)^v k} b^{v+w} \\
&= a^{i+(-1)^u (j+(-1)^v k)} b^{v+w} \\
&= a^{i+(-1)^u j+(-1)^{u+v} k} b^{u+v+w}.
\end{aligned}$$

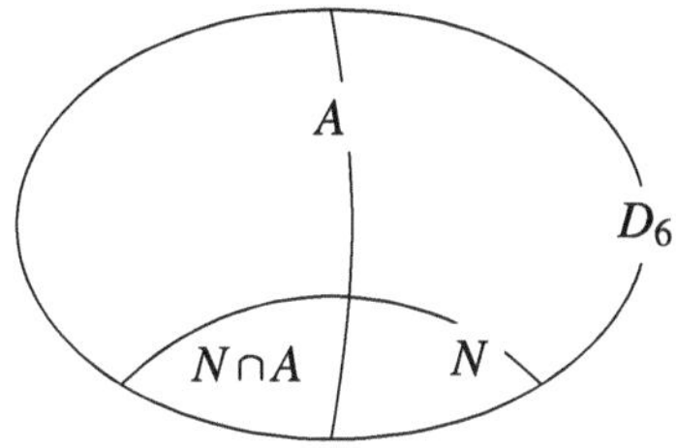

Fig. 3.6 $N \cap A$

¶ *Subgroups of D_6*

We immediately find a cyclic subgroup $A = \{e, a, \ldots, a^5\}$. Subgroups of A are enumerated as

$$A_1 = \{e, a^3\}, \quad A_2 = \{e, a^2, a^4\}, \quad \{e\}, \quad \text{and} \quad A,$$

which are also subgroups of D_6.

Let N be a subgroup of D_6 other than the subgroups of A. Theorem 3.10 claims that the intersection $N \cap A$ is a subgroup of D_6. Since $N \cap A \subset A$, the group $N \cap A$ is also a subgroup of A. See Fig. 3.6.
Since $N \cap A$ is a subgroup of A, the possibilities are:

$$N \cap A = \{e\}, \quad A_1, \quad A_2, \quad \text{or} \quad A.$$

(i) Suppose $N \cap A = \{e\}$. See Fig. 3.7. Take any two elements $a^m b, a^n b \in N - A$. Then, we have

$$\begin{aligned} (a^m b)(a^n b) &= a^m (ba^n) b \\ &= a^m (a^{6-n} b) b = a^{6+m-n} b^2 \\ &= a^{m-n}. \end{aligned}$$

Since $a^{m-n} \in A$ and $a^{m-n} = (a^m b)(a^n b) \in N$, we have $a^{m-n} \in N \cap A$, thus $a^{m-n} = e$. This means $m - n \equiv 0 \mod 6$, therefore $a^m b = a^n b$. We conclude that possibilities are limited to six cases:

$$N = \{e, a^n b\} \quad (n = 0, 1, \cdots, 5).$$

Quiz 3.13 Show that all the candidates of N above are subgroups of D_6.

(ii) Suppose $N \cap A = A_1$. See Fig. 3.8. Take any two elements $a^m b, a^n b \in N - A$. Then, we have

Fig. 3.7 $N \cap A = \{e\}$

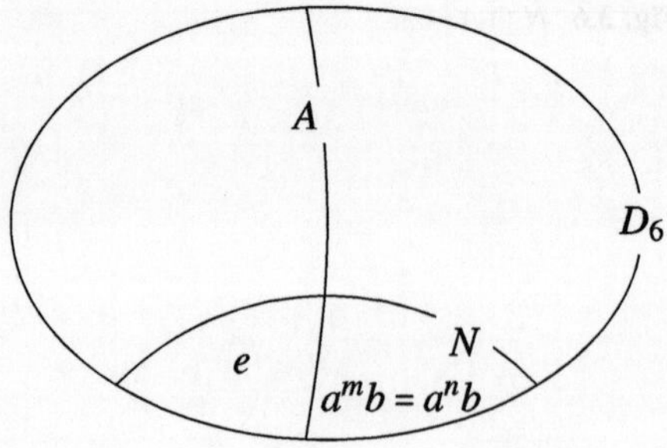

Fig. 3.8 $N \cap A = \{e, a^3\}$

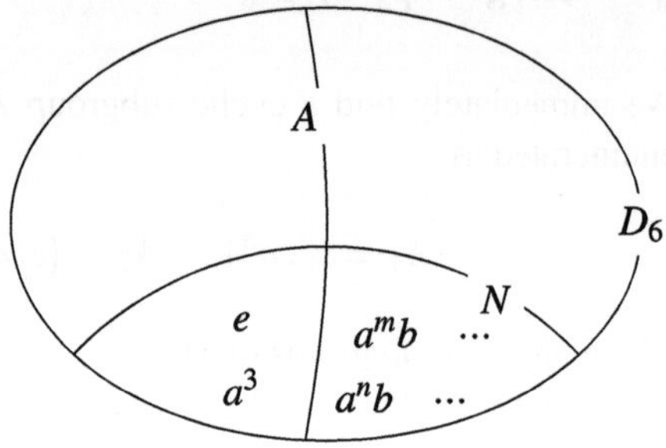

$$a^{m-n} \in N \cap A = A_1,$$

which means that $m - n \equiv 0 \mod 3$. The possibilities of N are three cases:

$$\{e, a^3, b, a^3 b\},$$
$$\{e, a^3, ab, a^4 b\}, \quad \text{and}$$
$$\{e, a^3, a^2 b, a^5 b\}.$$

Quiz 3.14 Show that all the candidates of N above are subgroups of D_6 making a multiplication table for each.

(iii) Suppose $N \cap A = A_2$. See Fig. 3.9. Take any two elements $a^m b, a^n b \in N - A$. Then, we have

$$a^{m-n} \in N \cap A = A_2.$$

Thus, we have $m - n \equiv 0 \mod 2$. There are two possibilities for N:

$$\{e, a^2, a^4, b, a^2 b, a^4 b\},$$
$$\{e, a^2, a^4, ab, a^3 b, a^5 b\}.$$

In the same way so far, these are clearly the subgroups of D_6.

Fig. 3.9 $N \cap A = \{e, a^2, a^4\}$

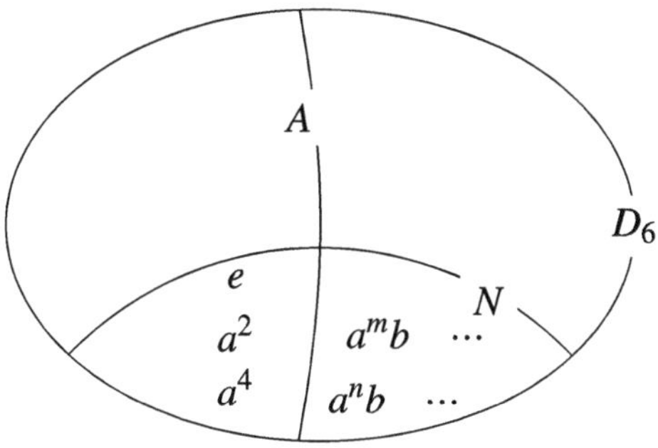

(iv) Suppose $N \cap A = A$. The size of a subgroup is a divisor of the size of the parent group. So, there are two possibilities, $N = A$ or $N = D_6$.

We have found all the subgroups of D_6. Next, we examine if each subgroup is normal or not.

Both the left and right $\{e\}$-orbits are singleton sets consisting of each element of D_6. So, $\{e\}$ is normal. Both the left and right D_6-orbits are identical to D_6. So, we see the D_6 is normal.

The size of the subgroup $A = \{e, a, \ldots, a^5\}$ is exactly the half of $|G|$. Since A is automatically an orbit, another half must be both left and right A-orbits. In general, if the size of a subgroup is the half of its parent group, then it is a normal subgroup. Thus, other two subgroups of size 6 are all normal.

Let us consider $A_1 = \{e, a^3\}$. The left A_1-orbit decomposition is given by

$$\begin{aligned} A_1 \backslash D_6 &= \{A_1, A_1 a, A_1 a^2, A_1 b, A_1 ab, A_1 a^2 b\} \\ &= \{\{e, a^3\}, \{a, a^4\}, \{a^2, a^5\}, \{b, a^3 b\}, \{ab, a^4 b\}, \{a^2 b, a^5 b\}\}. \end{aligned}$$

The right decomposition is

$$\begin{aligned} D_6 / A_1 &= \{A_1, a A_1, a^2 A_1, b A_1, ab A_1, a^2 b A_1\} \\ &= \{\{e, a^3\}, \{a, a^4\}, \{a^2, a^5\}, \{b, a^3 b\}, \{ab, a^4 b\}, \{a^2 b, a^5 b\}\}, \end{aligned}$$

which coincides with the left one. So, we conclude that A_1 is normal.

Consider $A_2 = \{e, a^2, a^4\}$. The left orbit decomposition is

$$\begin{aligned} A_2 \backslash D_6 &= \{A_2, A_2 a, A_2 b, A_2 ab\} \\ &= \{\{e, a^2, a^4\}, \{a, a^3, a^5\}, \{b, a^2 b, a^4 b\}, \{ab, a^3 b, a^5 b\}\}, \end{aligned}$$

while the right orbit decomposition is

$$\begin{aligned} D_6 / A_2 &= \{A_2, a A_2, b A_2, ab A_2\} \\ &= \{\{e, a^2, a^4\}, \{a, a^3, a^5\}, \{b, a^4 b, a^2 b\}, \{ab, a^5 b, a^3 b\}\}. \end{aligned}$$

Disregarding the order of elements in an orbit, they are identical. So, A_2 is a normal subgroup of D_6.

Consider $\{e, b\}$. The left orbit starting at a is $\{a, a^5b\}$, and the right orbit $\{a, ab\}$. So, we conclude that $\{e, b\}$ is not normal.

Quiz 3.15 Show that subgroups $\{e, ab\}$, $\{e, a^2b\}$, $\{e, a^3b\}$, $\{e, a^4b\}$, and $\{e, a^5b\}$ are not normal. Compare the left and right orbits starting at a.

Set $B = \{e, a^3, b, a^3b\}$. Since $Ba = \{a, a^4, a^5b, a^2b\}$ and $aB = \{a, a^4, ab, a^4b\}$ differ, B is not normal.

Quiz 3.16 Show that neither $\{e, a^3, ab, a^4b\}$ nor $\{e, a^3, a^2b, a^5b\}$ is normal. Compare the left and right orbit starting at a.

Exercises

[1] Consider the cyclic group $C_{12} = \{e, a, a^2, \cdots, a^{11}\}$ with rules $a^{12} = e$ and $a^m a^n = a^{m+n}$. An obvious generator is a.

(1) Calculate the order of each element of C_{12}.
(2) Enumerate all generators of C_{12}.
(3) For each $x \in C_{12}$ which is not a generator, find the subgroup generated by x:

$$\{e, x, x^2, \cdots\},$$

where $\cdots$ terminates in a finite number of steps.
(4) Let N be a non-trivial subgroup of C_{12}, namely $\{e\} \subsetneqq N \subsetneqq C_{12}$. Let k be the minimum of the set

$$\{n \mid 2 \leq n \leq 11, a^n \in N\}.$$

Show that all the elements of N can be expressed as

$$(a^k)^m = a^{km},$$

for some natural number m. We see that any subgroups of C_{12} is cyclic. Therefore, (3) exhausts all the subgroups of C_{12}.
(5) Find the orbit decomposition for each subgroup of C_{12}.

[2] Consider the infinite cyclic group $C_\infty = \{\cdots, a^{-2}, a^{-1}, e, a, a^2, \cdots\}$.

(1) Show that there are only two generators a and a^{-1}.
(2) Do the same investigations as in Exercise [1].
(3) Summarize the findings about the subgroups of C_∞.

[3] Consider the group $D_4 = \{e, a, a^2, a^3, b, ab, a^2b, a^3b\}$ with the following rules: $ba = a^3b, \quad ba^2 = a^2b, \quad ba^3 = ab,$ and $a^4 = b^2 = e.$

(1) Make a multiplication table of D_4. Confirm that it is really a group.
(2) Find all the subgroups of D_4.
(3) For each subgroup, decide if it is a normal subgroup or not.

[4] Consider the group $D_5 = \{e, a, a^2, a^3, a^4, b, ab, a^2b, a^3b, a^4b\}$ with the following rules: $ba^n = a^{5-n}b$ and $a^5 = b^2 = e.$

(1) Make a multiplication table of D_5. Confirm that it is really a group.
(2) Find all the subgroups of D_5.
(3) For each subgroup, decide if it is a normal subgroup or not.

[5] Consider the group $Q_8 = \{1, -1, i, -i, j, -j, k, -k\}$ with the following rules: $i^2 = j^2 = k^2 = -1, \quad ij = -ji = k, \quad jk = -kj = i,$ and $ki = -ik = j.$

(1) Make a multiplication table of Q_8. Confirm that it is really a group.
(2) Find all the subgroups of Q_8.
(3) For each subgroup, decide if it is a normal subgroup or not.

[6] Let $GL(n, \mathbb{R})$ and $GL(n, \mathbb{C})$ be the set of all $n \times n$ regular matrices with entries of real and complex numbers, respectively. Both are groups under matrix multiplication.

(1) Confirm that they are groups.
(2) Let $SL(n, \mathbb{C})$, $SL(n, \mathbb{R})$, and $SL(n, \mathbb{Z})$ be the groups of matrices with determinant 1 whose entries are taken from the indicated domains. Show that $SL(n, \mathbb{C})$ is a subgroup of $GL(n, \mathbb{C})$.
(3) Show that $SL(n, \mathbb{Z})$ is a subgroup of $SL(n, \mathbb{C})$.

Chapter 4
Homomorphisms of Groups

We study various properties of mappings between groups that preserve binary operations. Such mappings are called homomorphisms. Under a homomorphism, orbits of a normal subgroup are shown to behave *in clusters*.

4.1 Mappings that Preserve Binary Operations

We first make clear what it means to preserve binary operations.

Definition 4.1 A mapping $\varphi : G_1 \to G_2$ between groups G_1 and G_2 is called a *homomorphism* if

$$\varphi(xy) = \varphi(x)\varphi(y)$$

holds for any $x, y \in G_1$.

Example 4.2 Let $\mathbb{R}^+$ be the group of real numbers under addition. Let $\mathbb{R}_+^\times$ be the group of positive real numbers under multiplication. The mapping

$$\log : \mathbb{R}_+^\times \to \mathbb{R}^+,$$

is a homomorphism since

$$\log(xy) = \log x + \log y.$$

The exponential function is its inverse, which is also a homomorphism.

Example 4.3 We define the mapping $\varphi : \mathbb{R}^+ \to GL(2, \mathbb{R})$ as

S. Yukita, *First Lectures in Algebra*, Springer Asia Pacific Mathematics Series 10,
https://doi.org/10.1007/978-981-95-5457-7_4

$$\varphi : \theta \mapsto \begin{pmatrix} \cos\theta & -\sin\theta \\ \sin\theta & \cos\theta \end{pmatrix}.$$

This φ is a homomorphism. In fact, for any $\theta_1, \theta_2 \in \mathbb{R}^+$ we have

$$\begin{aligned} \varphi(\theta_1 + \theta_2) &= \begin{pmatrix} \cos(\theta_1+\theta_2) & -\sin(\theta_1+\theta_2) \\ \sin(\theta_1+\theta_2) & \cos(\theta_1+\theta_2) \end{pmatrix} \\ &= \begin{pmatrix} \cos\theta_1 & -\sin\theta_1 \\ \sin\theta_1 & \cos\theta_1 \end{pmatrix} \begin{pmatrix} \cos\theta_2 & -\sin\theta_2 \\ \sin\theta_2 & \cos\theta_2 \end{pmatrix} \\ &= \varphi(\theta_1)\varphi(\theta_2). \end{aligned}$$

Example 4.4 Let $\mathbb{C}^\times$ be the group of non-zero complex numbers under multiplication. The mapping $\det : GL(n, \mathbb{C}) \to \mathbb{C}^\times$ is a homomorphism. In fact, for any $P, Q \in GL(n, \mathbb{C})$ we have

$$\det(PQ) = \det P \times \det Q.$$

Example 4.5 Let $C_2 = \{e, a\}$ be a group with a rule $a^2 = e$. We define the mapping $\varphi : S_3 \to C_2 = \{e, a\}$ as

$$e,\ \sigma,\ \sigma^2 \mapsto e \qquad \tau,\ \sigma\tau,\ \sigma^2\tau \mapsto a.$$

To show φ is a homomorphism in a brute-force way, we have to show

$$\varphi(xy) = \varphi(x)\varphi(y)$$

for all $(x, y) \in S_3 \times S_3$. The number of such ordered pairs is 36.

If we use basic facts about homomorphisms, we can streamline the argument and avoid such a tedious brute-force attempt. We can easily find the following facts. Let $e_1 \in G_1$ and $e_2 \in G_2$ be the neutral elements.

(1) The neutral element is sent to the neutral element, namely $\varphi(e_1) = e_2$.

Proof Let $x \in G$ is an arbitrary element. Since φ is a homomorphism, we have $\varphi(e_1 x) = \varphi(e_1)\varphi(x)$. Using $e_1 x = x$, we have

$$\varphi(e_1)\varphi(x) = \varphi(x).$$

Multiplying both sides by $\varphi(x)^{-1}$ from the right, we have

$$\varphi(e_1) = \varphi(x)\varphi(x)^{-1} = e_2.$$

□

(2) An inverse is sent to an inverse, namely $\varphi(x)^{-1} = \varphi(x^{-1})$.

Proof Since φ is a homomorphism, we have $\varphi(x^{-1})\varphi(x) = \varphi(x^{-1}x) = \varphi(e_1)$. From (1) we have $\varphi(e_1) = e_2$. So, we have

$$\varphi(x^{-1})\varphi(x) = e_2,$$

which means $\varphi(x)^{-1} = \varphi(x^{-1})$. □

(3) If the order of $x \in G_1$ is finite, the order of $\varphi(x)$ divides that of x.

Proof Let n be the order of x. Then,

$$(\varphi(x))^n = \varphi(x^n) = \varphi(e_1) = e_2.$$

□

Quiz 4.1 Let x be an element of a group G with $x^n = e$ for some natural number n. Show that the order of x is a divisor of n.

Definition 4.6 Let $\varphi : G_1 \to G_2$ be a homomorphism between groups. The set of all $x \in G_1$ such that $\varphi(x) = e_2$ is called the *kernel* of φ and denoted by $\ker\varphi$. The image of φ is denoted by $\operatorname{Im}\varphi$. With a standard set notation, we can write them down as follows:

$$\ker\varphi \overset{\text{def}}{=} \{x \in G_1 \mid \varphi(x) = e_2\},$$

$$\operatorname{Im}\varphi \overset{\text{def}}{=} \{y \in G_2 \mid y = \varphi(x) \text{ for some } x \in G_1\}.$$

See Fig. 4.1.

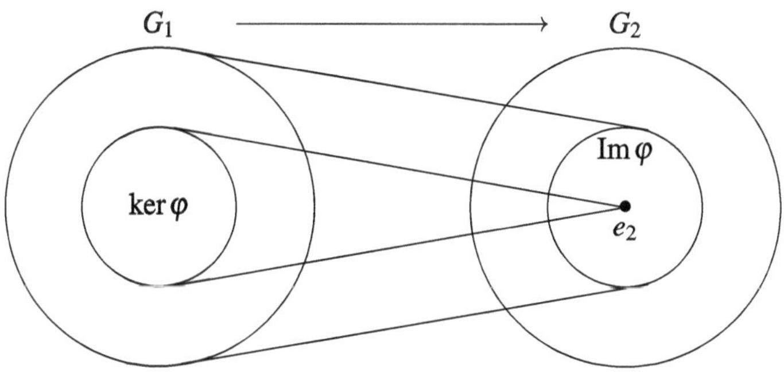

Fig. 4.1 Kernel and image

(4) The subset $\ker\varphi \subset G_1$ is a subgroup of G_1.

Proof Let x and y be arbitrary elements of $\ker\varphi$. Since $\varphi(x) = \varphi(y) = e_2$, we have

$$\varphi(xy) = \varphi(x)\varphi(y) = e_2e_2 = e_2,$$

which means that $\ker\varphi$ is closed under multiplication. From (2) we have

$$\varphi(x^{-1}) = \varphi(x)^{-1} = e_2^{-1} = e_2,$$

which means that $\ker\varphi$ is closed under inversion. □

(5) The subset $\operatorname{Im}\varphi \subset G_2$ is a subgroup of G_2.

Proof Let s and t be arbitrary elements of $\operatorname{Im}\varphi$. There are some elements $x, y \in G_1$ such that $s = \varphi(x)$ and $t = \varphi(y)$. Then, we have

$$st = \varphi(x)\varphi(y) = \varphi(xy),$$

which shows that $st \in \operatorname{Im}\varphi$ and that $\operatorname{Im}\varphi$ is closed under multiplication. From (2) we have

$$s^{-1} = \varphi(x)^{-1} = \varphi(x^{-1}),$$

which shows that $s^{-1} \in \operatorname{Im}\varphi$ and that $\operatorname{Im}\varphi$ is closed under inversion. □

In the next section, we will investigate the behavior of $(\ker\varphi)$-orbits under homomorphism.

4.2 Kernel Orbits Behave in Clusters

Given a homomorphism $\varphi : G_1 \to G_2$, we know that $\ker\varphi$ a subgroup of G_1. In this section, we will see that $\ker\varphi$ is a normal subgroup of G_1. For this purpose, we first investigate the behavior of $(\ker\varphi)$-orbits under homomorphism φ.

Lemma 4.7 *Given a homomorphism $\varphi : G_1 \to G_2$ between two groups G_1 and G_2, the left $(\ker\varphi)$-orbits behave in clusters and in a mutually exclusive way under the homomorphism φ. More precisely, given two different left $(\ker\varphi)$-orbits $(\ker\varphi)x$ and $(\ker\varphi)y$, we have*

$$\{\varphi(x)\} = \varphi((\ker\varphi)x) \stackrel{\text{def}}{=} \{\varphi(t) \mid t \in (\ker\varphi)x\},$$

$$\{\varphi(y)\} = \varphi((\ker\varphi)y) \stackrel{\text{def}}{=} \{\varphi(t) \mid t \in (\ker\varphi)y\}, \quad \textit{and}$$

$$\varphi(x) \neq \varphi(y).$$

Proof We first show that $\varphi((\ker\varphi)x) = \{\varphi(x)\}$. For any $s \in \ker\varphi$ we have

$$\varphi(sx) = \varphi(s)\varphi(x) = e_2\varphi(x) = \varphi(x),$$

which shows that all the elements of the left $(\ker\varphi)$-orbit starting at x are sent to the same element $\varphi(x) \in G_2$.

Let $(\ker\varphi)y$ be another left $(\ker\varphi)$-orbit. All the elements on this orbit are sent to a single element $\varphi(y)$. Suppose $\varphi(x) = \varphi(y)$, namely

$$\varphi(x)\varphi(y^{-1}) = e_2.$$

The LHS is

$$\varphi(xy^{-1}),$$

therefore $xy^{-1} \in \ker\varphi$. This gives

$$(\ker\varphi)x = (\ker\varphi)y,$$

namely the same left orbit. □

We can repeat the same discussion, replacing the left orbits by the right orbits.

Lemma 4.8 *Given a homomorphism $\varphi : G_1 \to G_2$ between two groups G_1 and G_2, the right $(\ker\varphi)$-orbits behave in clusters and mutually exclusively under the homomorphism φ. More precisely, given two different right $(\ker\varphi)$-orbits $x(\ker\varphi)$ and $y(\ker\varphi)$, we have*

$$\{\varphi(x)\} = \varphi(x(\ker\varphi)) \stackrel{\text{def}}{=} \{\varphi(t) \mid t \in x(\ker\varphi)\},$$
$$\{\varphi(y)\} = \varphi(y(\ker\varphi)) \stackrel{\text{def}}{=} \{\varphi(t) \mid t \in y(\ker\varphi)\}, \quad \textit{and}$$
$$\varphi(x) \neq \varphi(y).$$

Quiz 4.2 Write down a proof.

We introduce a new technical term for ease of formulating "behavior in clusters."

Definition 4.9 Let $\varphi : A \to B$ a mapping. For any $b \in \operatorname{Im}\varphi$, the preimage of $\{b\}$, namely

$$\varphi^{-1}(b) \stackrel{\text{def}}{=} \{a \in A \mid \varphi(a) = b\}$$

is called the *fiber* of b under the mapping φ. See Fig. 4.2.

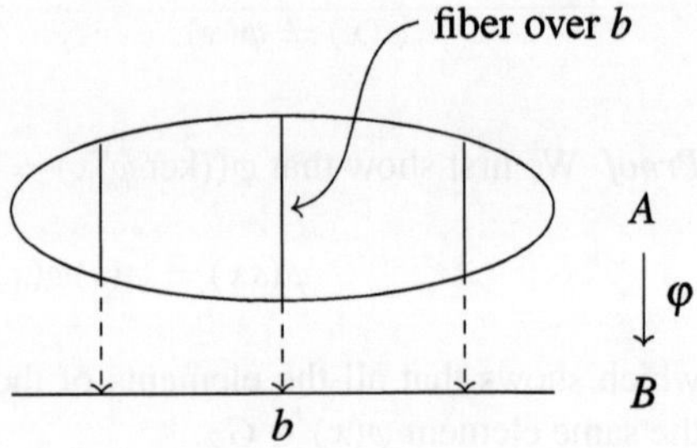

Fig. 4.2 Fibers of φ

Remark 4.10 A fiber may be just called a *fiber* of φ. Each fiber behaves in a cluster under the mapping φ.

Theorem 4.11 *Let $\varphi : G_1 \to G_2$ be a homomorphism of groups. Every fiber of φ is a left ($\ker\varphi$)-orbit and right ($\ker\varphi$)-orbit at the same time. This means that the left and right ($\ker\varphi$)-orbits are identical for any $x \in G_1$. See Fig. 4.3.*

Proof From Lemmas 4.7 and 4.8, we see that the left and right orbits coincide. □

We immediately have the following theorem.

Theorem 4.12 *Let $\varphi : G_1 \to G_2$ be a group homomorphism. The subgroup* $\ker\varphi$ *is a normal subgroup of G_1.*

Let us forget about homomorphisms for a while and concentrate on orbits of normal subgroups.

Let H be a normal subgroup of G. We know $Hx = xH$ For any $x \in G$. Therefore, we can simply say H-orbit without distinguishing left or right.

We can introduce a multiplication • between H-orbits. Let us see if the following definition makes sense. For any $xH, yH \in G/H$, set

$$(xH) \bullet (yH) \overset{\text{def}}{=} (xy)H.$$

To show that this is *well-defined*, we must establish the following.

Proposition 4.13 *Let x' and y' be other possible representatives of xH and yH, respectively. Then, we have $(x'y')H = (xy)H$.*

If this is true, replacing representatives with their alternatives on the same orbits does not alter the result of multiplication, which is the well-definedness of • as a binary operation between orbits.

Proof Since

$$x' \in xH, \quad y' \in yH,$$

there are $h_1, h_2 \in H$ such that $x' = xh_1$ and $y' = yh_2$. Thus, we have

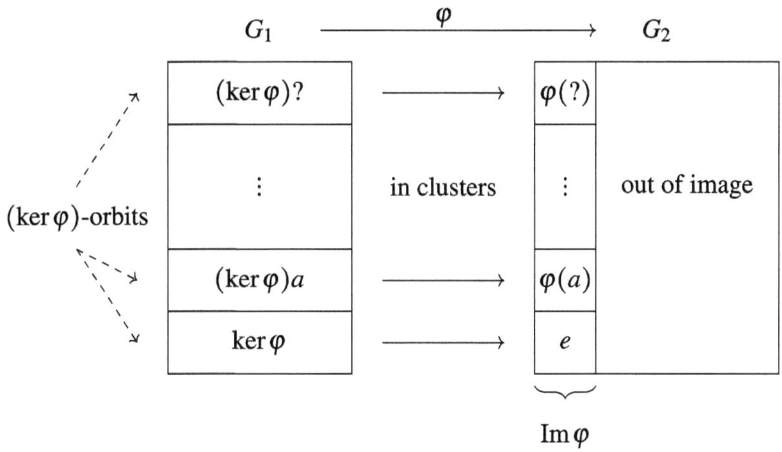

Fig. 4.3 Every fiber is a $(\ker\varphi)$-orbit

$$x'y' = xh_1yh_2.$$

Since $h_1y \in Hy = yH$, there is $h_3 \in H$ such that $h_1y = yh_3$. Thus, we have

$$x'y' = xyh_3h_2 \in xyH,$$

which means $x'y'H = xyH$. □

G/H is a group under $\bullet$. The neutral element is $eH = H$. The inverse of xH is $x^{-1}H$. Associativity comes from G.

Definition 4.14 Let H be a normal subgroup of G. If we define a binary operation $\bullet$ on G/H as

$$(xH) \bullet (yH) = xyH,$$

then G/H is a group under $\bullet$. The group G/H is called a *factor group* or *quotient group*.

Theorem 4.15 *Let H be a normal subgroup of G. The map φ defined by*

$$\begin{aligned} \varphi : G &\to G/H \\ x &\mapsto xH \end{aligned}$$

is a surjective homomorphism. We have $\ker\varphi = H$.

Proof It is clear that φ is surjective. We see that the map φ preserves multiplication as follows. By Proposition 4.13, we have

$$\varphi(xy) = xyH = (xH)(yH) = \varphi(x)\varphi(y).$$

Clearly, we have $\ker\varphi \supset H$. Let us show the converse inclusion. For any $x \in \ker\varphi$, we have $\varphi(x) = xH = H$. This means $x \in H$. Thus, $\ker\varphi \subset H$. We conclude that $\ker\varphi = H$. □

Definition 4.16 The homomorphism $\varphi : G \to G/H$ in Theorem 4.15 is called the *canonical projection map* or *canonical projection homomorphism.*

Here is the summary of this section.

- Let $\varphi : G_1 \to G_2$ be a group homomorphism. All the fibers of φ are a $(\ker\varphi)$-orbit. The mapping φ sends any orbit as a whole to an element of G_2. Different orbits are sent to different elements of G_2.
- Let H be a normal subgroup of G. Then, G/H is a group under the multiplication between orbits. There is the canonical projection map $G \to G/H$, whose kernel is H.
- If H is a subgroup of G which is not normal, namely $G/H \neq H\backslash G$, then G/H has no chance of being a group in a natural way.
- Any normal subgroup is a kernel of some homomorphism, for example the canonical projection map.

Example 4.17 Notice how the situations differ in the case of a normal subgroup and in the case of not normal. We examine subgroups $A = \{e, \sigma, \sigma^2\}$ and $B = \{e, \tau\}$ of

$$S_3 = \{e, \sigma, \sigma^2, \tau, \sigma\tau, \sigma^2\tau\}.$$

Recall that A is normal and B is not normal.

We want to rearrange the multiplication table of S_3 in a way that orbits are placed in clusters. Table 4.1 consists of four 3×3 blocks, each of which is an A-orbit. Different A-orbits are completely separated. This shows that multiplication among A-orbits fits nicely in the S_3 table. In Table 4.2 the title row and column are clustered in the right B-orbits. However, most of the 2×2 blocks are mixtures of orbits. Take, for example, the 2×2 block at the bottom left

$$\{\sigma^2, \sigma^2\tau\} \times \{\sigma^2, \sigma^2\tau\},$$

whose entries are

$$\sigma, \quad \sigma\tau, \quad \tau, \quad \text{and} \quad e.$$

Taking the left B-orbits as clusters will lead to the similar result.

Quiz 4.3 Using the left B-orbits, rearrange the S_3 table to get one similar to Table 4.2. Then, observe the mixture of orbits.

Table 4.1 A-orbits behave in clusters

	e	σ	σ^2	τ	$\sigma\tau$	$\sigma^2\tau$
e	e	σ	σ^2	τ	$\sigma\tau$	$\sigma^2\tau$
σ	σ	σ^2	e	$\sigma\tau$	$\sigma^2\tau$	τ
σ^2	σ^2	e	σ	$\sigma^2\tau$	τ	$\sigma\tau$
τ	τ	$\sigma^2\tau$	$\sigma\tau$	e	σ^2	σ
$\sigma\tau$	$\sigma\tau$	τ	$\sigma^2\tau$	σ	e	σ^2
$\sigma^2\tau$	$\sigma^2\tau$	$\sigma\tau$	τ	σ^2	σ	e

Table 4.2 Right B-orbits do not behave in clusters

	e	τ	σ	$\sigma\tau$	σ^2	$\sigma^2\tau$
e	e	τ	σ	$\sigma\tau$	σ^2	$\sigma^2\tau$
τ	σ	e	$\sigma^2\tau$	σ^2	$\sigma\tau$	σ
σ	σ	$\sigma\tau$	σ^2	$\sigma^2\tau$	e	τ
$\sigma\tau$	$\sigma\tau$	σ	τ	e	$\sigma^2\tau$	σ^2
σ^2	σ^2	$\sigma^2\tau$	e	τ	σ	$\sigma\tau$
$\sigma^2\tau$	$\sigma^2\tau$	σ^2	$\sigma\tau$	σ	τ	e

Exercises

[1] Let C_2, C_3, C_4, and C_6 be cyclic groups with generators a, b, c, and d, respectively. Use the same symbol e for the neutral elements of all groups. Find all homomorphisms between each pair of cyclic groups below, noting that the image of a generator decides everything of a homomorphism. Pay attention to the order of a candidate for an image of the generator. For each homomorphism, write down its ker and Im explicitly.

(1) $C_2 \to C_4$.
(2) $C_4 \to C_2$.
(3) $C_3 \to C_6$.
(4) $C_3 \to C_4$.

[2] Try the same thing as Exercise [1]. This time the codomain is S_3.

(1) $C_2 \to S_3$.
(2) $C_3 \to S_3$.
(3) $C_6 \to S_3$.

[3] Enumerate all normal subgroups of D_4 and Q_8 (see exercises of Chap. 3). Write down the multiplication table of each quotient group obtained from them.

Table 4.1 [illegible] orbits behave in clusters

[illegible]

Table 4.2 [illegible] orbits do not behave in clusters

[illegible]

Chapter 5
Symmetric Groups

We have studied the group S_3 in depth. In this chapter, we first discuss the symmetric group S_n of degree n in general. Then, we embark on an in-depth study of S_4.

5.1 Symmetric Group S_n

We introduce the concepts of conjugate elements, conjugate classes, and conjugate subgroups, and get a deeper insight into normal subgroups. The symmetric groups provide us with the most lucid examples to discuss conjugacies.

The group S_n is the group of all the permutations $\{1, 2, \cdots, n\}$, namely the group of all bijections $\{1, 2, \cdots, n\} \to \{1, 2, \cdots, n\}$. We immediately see $|S_n| = n!$. We denote an element $\sigma : \{1, 2, \cdots, n\} \to \{1, 2, \cdots, n\}$ of S_n by

$$\begin{pmatrix} 1 & 2 & \cdots & n \\ \sigma(1) & \sigma(2) & \cdots & \sigma(n) \end{pmatrix}.$$

For example

$$a = \begin{pmatrix} 1\,2\,3\,4\,5\,6\,7\,8\,9 \\ 3\,4\,7\,6\,2\,8\,1\,5\,9 \end{pmatrix} \quad \in S_9$$

denotes a bijection

$$\begin{matrix} 1 & 2 & 3 & 4 & 5 & 6 & 7 & 8 & 9 \\ \downarrow & \downarrow & \downarrow & \downarrow & \downarrow & \downarrow & \downarrow & \downarrow & \downarrow \\ 3 & 4 & 7 & 6 & 2 & 8 & 1 & 5 & 9 \end{matrix}.$$

S. Yukita, *First Lectures in Algebra*, Springer Asia Pacific Mathematics Series 10,
https://doi.org/10.1007/978-981-95-5457-7_5

We can visualize a in the following diagram:

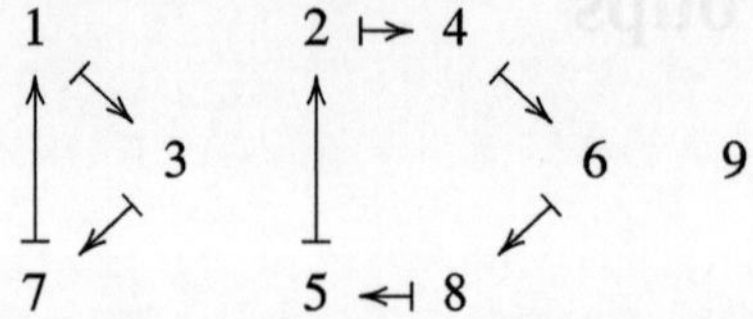

This diagram suggests that permutation a is decomposed as a product of cycles

$$a = (1\ 3\ 7)(2\ 4\ 6\ 8\ 5),$$

where 9 is omitted since it is a fixed point.

If two cycles $(m_1 m_2 \cdots m_k)$ and $(n_1 n_2 \cdots n_\ell)$ are disjoint, namely they have no elements in common, then they are commutative (we will formally prove this in Proposition 5.4). Therefore, we can safely change the order of cycles and express a as

$$a = (2\ 4\ 6\ 8\ 5)(1\ 3\ 7).$$

Let us proceed to formal definitions.

Definition 5.1 Let $\sigma \in S_n$ and $k \in \{1, \cdots, n\}$ such that:

- For some $i \in \{1, \cdots, n\}$,

$$i, \sigma(i), \sigma^2(i), \cdots, \quad \text{and} \quad \sigma^{\ell-1}(i)$$

  are mutually different and $\sigma^\ell(i) = i$. The diagram below may appeal to intuition:

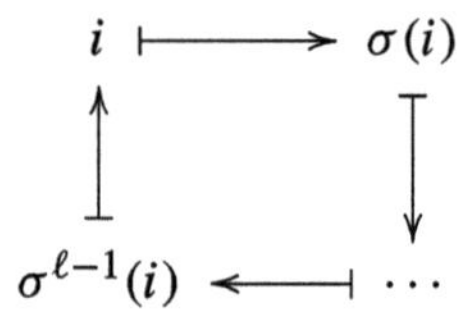

- For any k in

$$\{1, \cdots, n\} - \{i, \sigma(i), \sigma^2(i), \cdots, \sigma^{\ell-1}(i)\},$$

  we have $\sigma^\ell(k) = k$.

Then, the permutation σ is called a *cycle* of length ℓ and denoted by

$$\sigma = (i\ \sigma(i)\ \cdots\ \sigma^{\ell-1}(i)).$$

When a permutation is applied to an element in $\{1, \cdots, n\}$, say (1 2 3)(4 5) is applied to 2, we write

$$(1\ 2\ 3)(4\ 5)2 = 3.$$

Remark 5.2 The expression of a cycle is not unique. For example, we have

$$(1\ 2\ 3\ 4) = (2\ 3\ 4\ 1) = (3\ 4\ 1\ 2) = (4\ 1\ 2\ 3).$$

Definition 5.3 Two cycles $(m_1 m_2 \cdots m_k)$ and $(n_1 n_2 \cdots n_\ell)$ are said to be *disjoint* if

$$\{m_1, m_2, \cdots, m_k\} \cap \{n_1, n_2, \cdots, n_\ell\} = \emptyset.$$

Proposition 5.4 *If two cycles $(m_1 m_2 \cdots m_k)$ and $(n_1 n_2 \cdots n_\ell)$ are disjoint, then they are commutative:*

$$(m_1 m_2 \cdots m_k)(n_1 n_2 \cdots n_\ell) = (n_1 n_2 \cdots n_\ell)(m_1 m_2 \cdots m_k).$$

When we want to prove two mappings are the same, a brute-force way is to show the equality pointwise. This is one of the design patterns of proof which we will call the *pointwise pattern*. Let us apply this design pattern in the following proof.

Proof

(1) If $x \notin \{m_1, m_2, \cdots, m_k\} \cup \{n_1, n_2, \cdots, n_\ell\}$, then we have

$$(m_1 m_2 \cdots m_k)x = x \quad \text{and} \quad (n_1 n_2 \cdots n_\ell)x = x.$$

Thus, the following holds:

$$(m_1 m_2 \cdots m_k)(n_1 n_2 \cdots n_\ell)x = (n_1 n_2 \cdots n_\ell)(m_1 m_2 \cdots m_k)x = x.$$

(2) Let $x \in \{m_1, m_2, \cdots, m_k\}$. From the disjointness assumption we know $x \notin \{n_1, n_2, \cdots, n_\ell\}$, which means $(n_1 n_2 \cdots n_\ell)x = x$. Thus, we have

$$(m_1 m_2 \cdots m_k)(n_1 n_2 \cdots n_\ell)x = (m_1 m_2 \cdots m_k)x.$$

Since $(m_1 m_2 \cdots m_k)x \in \{m_1, m_2, \cdots, m_k\}$, disjointness gives $(m_1 m_2 \cdots m_k)x \notin \{n_1, n_2, \cdots, n_\ell\}$. Therefore, we have

$$(n_1 n_2 \cdots n_\ell)\,\underline{(m_1 m_2 \cdots m_k)x} = (m_1 m_2 \cdots m_k)x.$$

The two results coincide.

(3) If $x \in \{n_1, n_2, \cdots, n_\ell\}$, the argument goes exactly the same as in (2).

We have examined all possible cases. □

Quiz 5.1 Write down the details of (3) above.

Proposition 5.5 *Any permutation $\sigma \in S_n$ can be expressed as a product of mutually commutative cycles. The order of σ is the least common multiple of the lengths of such cycles.*

In order to show something exists, a brute-force way is to provide an algorithm to obtain the thing. We may call this way of reasoning the *concrete algorithm pattern*.

Proof We give an algorithm to obtain the product of mutually commutative cycles.

(1) We consider the sequence

$$1, \sigma(1), \sigma^2(1), \cdots .$$

Let k be the smallest number such that $\sigma^k(1) = 1$, which should surely exist since the order of σ is finite, more precisely a divisor of $n!$. All the elements in the sequence

$$1, \sigma(1), \sigma^2(1), \cdots, \sigma^{k-1}(1),$$

are mutually different. The choice of k ensures us $\sigma^i(1) \neq 1$ for all $1 < i < k$. Further, there is no pair (i, j) such that $1 < i < j < k$ and $\sigma^i(1) = \sigma^j(1)$. Otherwise we would have $\sigma^{j-i}(1) = 1$ and $0 < j - i < k$, which contradicts the choice of k. To sum up, the permutation σ acts on the subset $\{1, \sigma(1), \sigma^2(1), \cdots, \sigma^{k-1}(1)\}$ of $\{1, \cdots, n\}$ like this:

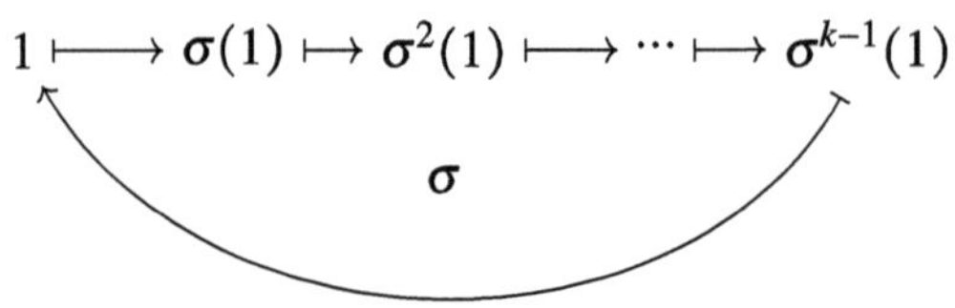

(2) Take the smallest number i from

$$\{1, \cdots, n\} - \{1, \sigma(1), \sigma^2(1), \cdots, \sigma^{k-1}(1)\}.$$

If the above set is empty, the permutation σ is already a cycle itself, and we have nothing to do after step (1). We consider the sequence

$$i, \sigma(i), \sigma^2(i), \cdots .$$

Let ℓ be the smallest number such that $\sigma^\ell(i) = i$. By the same discussion as in step (1), we know that the permutation σ acts on the subset $\{i, \sigma(i), \sigma^2(i), \cdots, \sigma^{\ell-1}(i)\}$ of $\{1, \cdots, n\}$ like this:

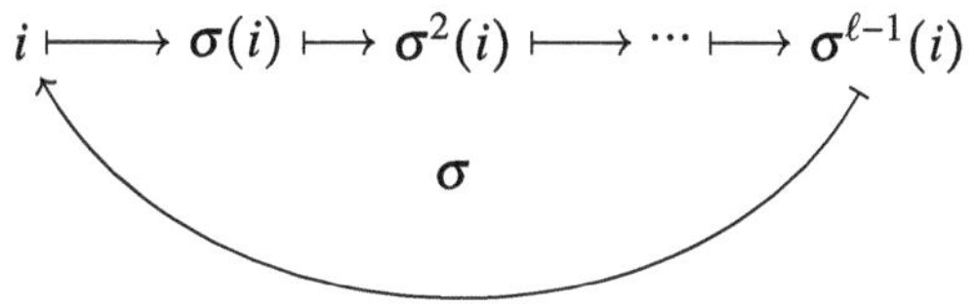

(3) Take the smallest number j from

$$\{1, \cdots, n\} - \{1, \sigma(1), \sigma^2(1), \cdots, \sigma^{k-1}(1)\} - \{i, \sigma(i), \sigma^2(i), \cdots, \sigma^{\ell-1}(i)\}.$$

If the above set is empty, the permutation σ is

$$(1\ \sigma(1)\ \cdots\ \sigma^{k-1}(1))(i\ \sigma(i)\ \cdots\ \sigma^{\ell-1}(i))$$

and we can conclude here.

(4) Such a procedure terminates within a finite number of iterations. The permutation σ is the product of so obtained cycles that are mutually commutative.

Then, we focus on the order of σ. Let $\sigma = c_1 c_2 \cdots c_k$ be the decomposition of mutually commutative cycles. The t-th power of σ is

$$\sigma^t = c_1^t c_2^t \cdots c_k^t.$$

In order for σ^t to be e, all $c_1^t, \cdots, c_k^t$ must become e at the same time, which means that the order of σ is the least common multiple of the orders (lengths) of cycles of $c_1, \cdots, c_k$. □

Remark 5.6 Given a product of cycles that are not commutative, its n-th power may not be the product of the n-th powers of the cycles. For example, in S_3, we have

$$((1\ 2\ 3)(1\ 2))^3 \neq (1\ 2\ 3)^3(1\ 2)^3,$$

where the LHS is (1 3) and the RHS is (1 2).

Quiz 5.2 Confirm the above inequality by calculation.

Let us see how the algorithm in Proposition 5.5 works with a concrete example.

Example 5.7 Consider an element $a \in S_9$

$$a = \begin{pmatrix} 1\,2\,3\,4\,5\,6\,7\,8\,9 \\ 3\,4\,7\,6\,2\,8\,1\,5\,9 \end{pmatrix}.$$

(1) The sequence beginning with 1 terminates as follows:

$$1, a(1) = 3, a(3) = 7, a(7) = 1.$$

So, we find a cycle $(1\ 3\ 7)$.

(2) The smallest number in

$$\{1, \cdots, 9\} - \{1, 3, 7\}$$

is 2. The sequence beginning with 2 terminates as follows:

$$2,\ a(2) = 4,\ a(4) = 6,\ a(6) = 8,\ a(8) = 5,\ a(5) = 2.$$

We find a cycle $(2\ 4\ 6\ 8\ 5)$.

(3) The set

$$\{1, \cdots, 9\} - \{1, 3, 7\} - \{2, 4, 6, 8, 5\}$$

contains only one element 9 which is a fixed point of a, which alone constitutes a cycle of length 1.

(4) The set

$$\{1, \cdots, 9\} - \{1, 3, 7\} - \{2, 4, 6, 8, 5\} - \{9\}$$

is empty. So, the procedure terminates here.

We have obtained

$$a = (1\ 3\ 7)(2\ 4\ 6\ 8\ 5),$$

omitting a cycle of length 1.

Definition 5.8 A cycle of length 2, $\sigma = (m\ n)$, is called especially a *transposition*.

We will see that transpositions generate everything.

Example 5.9 We have the following:

$$(1\ 2\ 3\ 4\ 5) = (1\ 5)(1\ 4)(1\ 3)(1\ 2).$$

We can prove this equality using the pointwise pattern as follows:

$$(1\ 5)(1\ 4)(1\ 3)(1\ 2)1 = 2,$$

$$(1\ 5)(1\ 4)(1\ 3)(1\ 2)2 = 3,$$

$$(1\ 5)(1\ 4)(1\ 3)(1\ 2)3 = 4,$$

$$(1\ 5)(1\ 4)(1\ 3)(1\ 2)4 = 5,$$
$$(1\ 5)(1\ 4)(1\ 3)(1\ 2)5 = 1.$$

Proposition 5.10 *Any cycle can be expressed as a product of transpositions:*

$$(a_1\ a_2\ \cdots\ a_n) = (a_1\ a_n)(a_1\ a_{n-1})\cdots(a_1\ a_2).$$

Proof We only have to show

$$(a_1 a_2 \cdots a_n) = (a_1 a_n)(a_1 \cdots a_{n-1}). \tag{5.1}$$

In fact, we can use this formula repeatedly to get the goal as follows:

$$\begin{aligned}
&(a_1\ a_2\ \cdots\ a_n)\\
=&(a_1 a_n)(a_1 \cdots a_{n-1})\\
=&(a_1 a_n)(a_1 a_{n-1})(a_1 \cdots a_{n-2})\\
=&(a_1 a_n)(a_1 a_{n-1})(a_1 a_{n-2})(a_1 \cdots a_{n-3})\\
&\vdots\\
=&(a_1\ a_n)(a_1\ a_{n-1})\cdots(a_1\ a_2).
\end{aligned}$$

We prove Eq. (5.1) using the pointwise pattern. Namely, we show that the equation

$$(a_1 a_2 \cdots a_n)a = (a_1 a_n)(a_1 \cdots a_{n-1})a$$

holds for all $a \in \{1, 2, \cdots, n\}$.

(1) If $a \neq \{a_1, a_2, \cdots, a_n\}$, then we have

$$(a_1\ a_2\ \cdots\ a_n)a = a$$
$$(a_1 a_n)(a_1 \cdots a_{n-1})a = a,$$

and therefore

$$(a_1\ a_2\ \cdots\ a_n)a = (a_1 a_n)(a_1 \cdots a_{n-1})a.$$

(2) If $a = a_i$ for some $1 \leq i \leq n-2$, then we have

$$(a_1\ a_2\ \cdots\ a_n)a_i = a_{i+1}$$
$$(a_1 a_n)(a_1 \cdots a_{n-1})a_i = (a_1 a_n)a_{i+1} = a_{i+1},$$

and therefore

$$(a_1\ a_2\ \cdots\ a_n)a_i = (a_1a_n)(a_1 \cdots a_{n-1})a_i.$$

(3) If $a = a_{n-1}$, then we have

$$(a_1\ a_2\ \cdots\ a_n)a_{n-1} = a_n$$
$$(a_1a_n)(a_1 \cdots a_{n-1})a_{n-1} = (a_1a_n)a_1 = a_n,$$

and therefore

$$(a_1\ a_2\ \cdots\ a_n)a_{n-1} = (a_1a_n)(a_1 \cdots a_{n-1})a_{n-1}.$$

(4) If $a = a_n$, then we have

$$(a_1\ a_2\ \cdots\ a_n)a_n = a_1$$
$$(a_1a_n)(a_1 \cdots a_{n-1})a_n = (a_1a_n)a_n = a_1,$$

and therefore

$$(a_1\ a_2\ \cdots\ a_n)a_n = (a_1a_n)(a_1 \cdots a_{n-1})a_n.$$

We have examined all possible cases for a. We conclude

$$(a_1a_2 \cdots a_n) = (a_1a_n)(a_1 \cdots a_{n-1}).$$

□

Theorem 5.11 *The symmetric group S_n is generated by the following set of transpositions:*

$$\{(1\ 2), (1\ 3), (1\ 4), \cdots, (1\ n)\}.$$

Proof By Proposition 5.4, we can express any $\sigma \in S_n$ as a product of disjoint cycles. By Proposition 5.10, we can express each cycle as a product of transpositions. Further, we can express any transposition $(i\ j)$ as a product of transpositions $(1\ i)$. For $1 < i < j$, we have

$$(i\ j) = (1\ j)(1\ i)(1\ j),$$

which we can easily prove using the pointwise pattern. We omit the details. We can illustrate the situation as follows: %TagUnImg

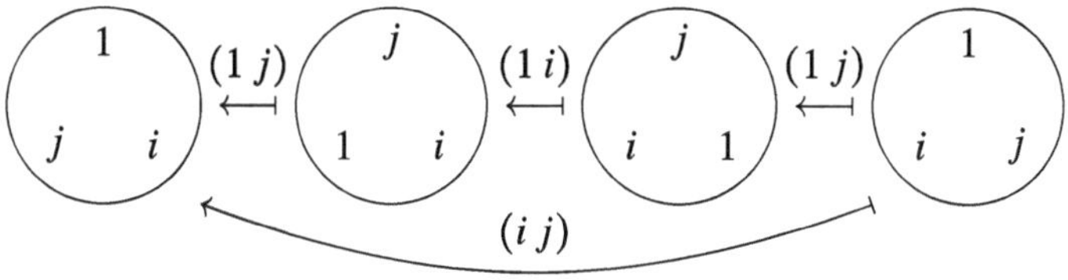

□

Remark 5.12 Notice that the transpositions appearing in the product above are not mutually commutative in general. We calculated the following repeatedly:

$$(1\ 3)(1\ 2) \neq (1\ 2)(1\ 3).$$

Theorem 5.13 *Given a permutation $\sigma \in S_n$, it can be expressed as a product of transpositions in many ways. Further, the numbers of transpositions appearing in such products may vary. However, the parity of the numbers, namely even or odd, is determined by σ.*

To prove this theorem, we usually use the Vandermonde polynomial.

Definition 5.14 A polynomial of n variables

$$\Delta(X_1, X_2, \cdots, X_n) \stackrel{\text{def}}{=} \prod_{1 \leq i < j \leq n} (X_i - X_j)$$

is called the *Vandermonde polynomial.*

Remark 5.15 The above product can be exhaustively expressed as follows:

$$\begin{array}{llll} (X_1 - X_2) & \times(X_1 - X_3) & \cdots & \times(X_1 - X_n) \\ & \times(X_2 - X_3) & \cdots & \times(X_2 - X_n) \\ & & \ddots & \vdots \\ & & & \times(X_{n-1} - X_n) \end{array}$$

Definition 5.16 Let $\mathbb{C}[X_1, \cdots, X_n]$ be the set of all polynomials with coefficients in $\mathbb{C}$ and variables $X_1, X_2, \cdots, X_n$. For any $\sigma \in S_n$ and $f(X_1, \cdots, X_n) \in \mathbb{C}[X_1, \cdots, X_n]$, let σ act on $f(X_1, \cdots, X_n)$ as

$$\sigma f(X_1, \cdots, X_n) \stackrel{\text{def}}{=} f(X_{\sigma(1)}, \cdots, X_{\sigma(n)}).$$

Example 5.17 Let $f(X_1, X_2, X_3) = X_1 + 2X_2 + 3X_3$ and $g(X_1, X_2, X_3) = (X_1 - X_2)(X_1 - X_3)(X_2 - X_3)$. We have the following:

$$\begin{aligned} (1\ 2\ 3) f(X_1, X_2, X_3) &= X_2 + 2X_3 + 3X_1 = 3X_1 + X_2 + 2X_3, \\ (1\ 2) f(X_1, X_2, X_3) &= X_2 + 2X_1 + 3X_3 = 2X_1 + X_2 + 3X_3, \end{aligned}$$

$$(1\ 2\ 3)g(X_1, X_2, X_3) = (X_2 - X_3)(X_2 - X_1)(X_3 - X_1) = g(X_1, X_2, X_3),$$
$$(1\ 2)g(X_1, X_2, X_3) = (X_2 - X_1)(X_2 - X_3)(X_1 - X_3) = -g(X_1, X_2, X_3).$$

Proposition 5.18 *The group S_n acts on the space $\mathbb{C}[X_1, \cdots, X_n]$ as a transformation group. Namely, for any $\sigma, \tau \in S_n$ and $f(X_1, \cdots, X_n) \in \mathbb{C}[X_1, \cdots, X_n]$, we have*

$$(\sigma\tau)f(X_1, \cdots, X_n) = \sigma(\tau f(X_1, \cdots, X_n)).$$

Proof By definition, the LHS is

$$\begin{aligned}(\sigma\tau)f(X_1, \cdots, X_n) &= f(X_{(\sigma\tau)1}, \cdots, X_{(\sigma\tau)n}) \\ &= f(X_{\sigma(\tau 1)}, \cdots, X_{\sigma(\tau n)}).\end{aligned}$$

By definition, the RHS is

$$\begin{aligned}\sigma(\tau f(X_1, \cdots, X_n)) &= \sigma f(X_{\tau 1}, \cdots, X_{\tau n}) \\ &= f(X_{\sigma(\tau 1)}, \cdots, X_{\sigma(\tau n)}).\end{aligned}$$

We see that they coincide. □

Lemma 5.19 *Let $\tau \in S_n$ be a transposition. Then, we have*

$$\tau\Delta(X_1, X_2, \cdots, X_n) = -\Delta(X_1, X_2, \cdots, X_n).$$

Proof Let $\tau = (i\ j)$ with $i < j$. We take indices k, ℓ, m from the intervals separated by i and j as in

$$1 \leq k < i < \ell < j < m \leq n.$$

(1) The factor $(X_i - X_j)$ is changed into $(X_j - X_i) = -(X_i - X_j)$ by τ. The sign of this factor is flipped.
(2) For $k < i$, the factors $(X_k - X_i)$ and $(X_k - X_j)$ are exchanged by τ. Neither of them changes its sign.
(3) For $i < \ell < j$, the factors $(X_i - X_\ell)$ and $(X_\ell - X_j)$ are exchanged by τ and flip their signs at the same time. So, $(X_i - X_\ell)(X_\ell - X_j)$ is preserved.
(4) For $j < m$, the factors $(X_i - X_m)$ and $(X_j - X_m)$ are exchanged by τ without touching their signs.
(5) Other factors are untouched by τ.

We have examined exhaustively the effects of τ factor by factor. Since (1) is the only case that causes the flip of the sign, the effect of τ on the Vandermonde polynomial is multiplying by -1. We may summarize all the correspondences in lexicographical order as follows, where k, ℓ, and m run through $1 \leq k < i$, $i < \ell < j$, and $j < m \leq n$ for fixed i and j:

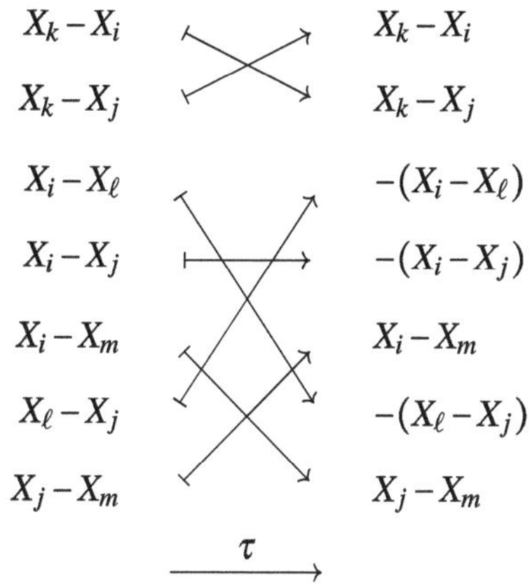

□

We return to Theorem 5.13.

Proof of Theorem 5.13 Let $\sigma \in S_n$. We can express σ as a product of, say k transpositions. By Proposition 5.18 and Lemma 5.19, we have

$$\sigma\Delta(X_1, X_2, \cdots, X_n) = (-1)^k \Delta(X_1, X_2, \cdots, X_n).$$

This result must be uniquely determined by σ. So, the parity of k is unique for each σ. □

Definition 5.20 If a permutation $\sigma \in S_n$ is expressed by an even number of transpositions, σ is called an *even permutation* and written $\text{sgn}(\sigma) = +1$. Otherwise, it is called an *odd permutation* and written $\text{sgn}(\sigma) = -1$.

The mapping $\text{sgn} : S_n \to \{\pm 1\}$ is a homomorphism from the group S_n to the group $\{\pm 1\}$. The set of all even permutations forms a subgroup of S_n, and it is the kernel of sgn, therefore a normal subgroup, whose size is the half of the parent's, namely $\frac{n!}{2}$.

Definition 5.21 The normal subgroup of S_n consisting of even permutations is called an n-th *alternating group* and denoted by A_n.

It is a good opportunity to review the concept of a normal subgroup and introduce additional concepts, which will be fully utilized in the next section where we undertake a deep study of S_4.

Let H be a normal subgroup of G. For any $a \in G$, we must have $aH = Ha$. Multiplying both sides by a^{-1} from the right, we get $aHa^{-1} = H$. We can reverse the process. Therefore, we can say that a subgroup H of G is a normal subgroup if and only if $aHa^{-1} = H$ for any $a \in G$.

If H is not a normal subgroup, then $aHa^{-1} \neq H$ for some $a \in G$. What kind of set is this? Is is still a subgroup of G or something else?

Proposition 5.22 *Let H be a subgroup of G. Take an arbitrary $a \in G$. Then, the subset aHa^{-1} is a subgroup of G.*

Proof Take any two elements $axa^{-1}, aya^{-1} \in aHa^{-1}$, where $x, y \in H$. If we multiply them, we get

$$axa^{-1}aya^{-1} = axya^{-1},$$

which is again an element of aHa^{-1} since $xy \in H$. Thus, the set aHa^{-1} is closed under multiplication. We can also find the inverse of any element in aHa^{-1}. Since

$$(axa^{-1})(ax^{-1}a^{-1}) = axx^{-1}a^{-1} = e,$$

we have

$$(axa^{-1})^{-1} = ax^{-1}a^{-1} \quad \in aHa^{-1}.$$

□

The concept of a normal subgroup can be given an alternative definition.

Definition 5.23 Let H be a subgroup of G. Taking any element $a \in G$, we say the subgroups H and aHa^{-1} are *conjugate subgroups* of G. We can also say aHa^{-1} is a *conjugate* of H in G. If a subgroup H has no conjugates other than itself, namely $aHa^{-1} = H$ for all $a \in G$, then H is called a *normal subgroup* of G.

Remark 5.24 The relation conjugate is an equivalence relation.

(1) Reflexive. A normal subgroup is a conjugate of itself. We see this since $H = eHe^{-1}$.
(2) Symmetric. If H_1 is a conjugate of H_2, then H_2 is a conjugate of H_1. In fact, if $H_1 = aH_2a^{-1}$, then $aH_2a^{-1} = (a^{-1})(aH_2a^{-1})(a^{-1})^{-1}$.
(3) Transitive. If H_1 is a conjugate of H_2, and H_2 is a conjugate of H_3, then H_1 is a conjugate of H_3. In fact, if $H_1 = aH_2a^{-1}$, $H_2 = bH_3b^{-1}$, then we have $H_1 = (ab)H_3(ab)^{-1}$.

The notion of conjugate is also applied to elements.

Definition 5.25 Let G be a group and $x, a \in G$. The elements x and axa^{-1} are called *conjugate elements*. We can also say that axa^{-1} is a *conjugate* of x.

The relation "conjugate" for elements is also an equivalence relation. We omit the proof.

Proposition 5.26 *Let H be a subgroup of G. If for every $x \in H$ its conjugates are all contained in H, then H is a normal subgroup of G. The converse is also true.*

Proof The "if" part of the statement is equivalent to say that $aHa^{-1} \subset H$ for all $a \in G$. Multiplying both sides of $\subset$ by a^{-1} from the left and by a from the right, we have $H \subset a^{-1}Ha$. These set inclusions hold for all $a \in G$, therefore we can replace a with a^{-1} at any points of the discussion. Now, we know $aHa^{-1} \subset H$ and $H \subset aHa^{-1}$, which leads to $aHa^{-1} = H$. This means that H is a normal subgroup of G.

The converse is trivial. □

In the group S_n, conjugate elements are characterized in the simplest and most beautiful way.

Let $\sigma, a \in S_n$ be expressed as a product of mutually disjoint cycles

$$\sigma = (c_1\ c_2\ \cdots\ c_\gamma)(d_1\ d_2\ \cdots\ d_\delta)\cdots(z_1\ z_2\ \cdots\ z_\omega).$$

Then, $a\sigma a^{-1}$ is expressed as a product of mutually disjoint cycles

$$(c_{a(1)}\ c_{a(2)}\ \cdots\ c_{a(\gamma)})(d_{a(1)}\ d_{a(2)}\ \cdots\ d_{a(\delta)})\cdots(z_{a(1)}\ z_{a(2)}\ \cdots\ z_{a(\omega)}).$$

Example 5.27 Let $\sigma, a \in S_8$ be

$$\sigma = (1\ 2\ 3)(4\ 5\ 6\ 7\ 8)$$

and

$$a = \begin{pmatrix} 1\,2\,3\,4\,5\,6\,7\,8 \\ 1\,3\,7\,2\,4\,6\,8\,5 \end{pmatrix}.$$

Then, we have

$$a\sigma a^{-1} = (1\ 3\ 7)(2\ 4\ 6\ 8\ 5).$$

Only with this single example, we are firmly convinced that two permutations are conjugate in S_n if and only if they have similar decompositions into the products of commuting cycles. Here, the word "similar" means that the lengths of cycles are equal as a whole.

5.2 Study of S_4

We discussed the symmetric groups S_n in general. In this section, we apply various techniques so far obtained to concrete examples taken from S_4. Children who failed to gain fluency in calculating elementary arithmetic would have difficulty in learning high-school algebra. This applies to the learning of group theory. So, let us improve, if necessary, our fluency in calculating permutations before embarking on the serious study.

Example 5.28 An efficient way of composing several permutations is to do it pointwise. Suppose we have to calculate $\sigma = (1\ 2)(1\ 2\ 3)(1\ 2)$.

- Tracing 1 as $(1\ 2)1 = 2,\ \ (1\ 2\ 3)2 = 3,\ \ (1\ 2)3 = 3$, we know $\sigma : 1 \mapsto 3$.
- Tracing 2 as $(1\ 2)2 = 1,\ \ (1\ 2\ 3)1 = 2,\ \ (1\ 2)2 = 1$, we know $\sigma : 2 \mapsto 1$.
- Tracing 3 as $(1\ 2)3 = 3,\ \ (1\ 2\ 3)3 = 1,\ \ (1\ 2)1 = 2$, we know $\sigma : 3 \mapsto 2$.

Gathering these pieces of information, we get

$$(1\ 2)(1\ 2\ 3)(1\ 2) = \begin{pmatrix} 1\ 2\ 3 \\ 3\ 1\ 2 \end{pmatrix} = (1\ 3\ 2).$$

An alternative strategy is to calculate by iterating binary composition from the right.

$$\begin{aligned} (1\ 2)(1\ 2\ 3)(1\ 2) &= \begin{pmatrix} 1\ 2\ 3 \\ 2\ 1\ 3 \end{pmatrix} \left(\begin{pmatrix} 1\ 2\ 3 \\ 2\ 3\ 1 \end{pmatrix} \begin{pmatrix} 1\ 2\ 3 \\ 2\ 1\ 3 \end{pmatrix} \right) \\ &= \begin{pmatrix} 1\ 2\ 3 \\ 2\ 1\ 3 \end{pmatrix} \begin{pmatrix} 1\ 2\ 3 \\ 3\ 2\ 1 \end{pmatrix} \\ &= \begin{pmatrix} 1\ 2\ 3 \\ 3\ 1\ 2 \end{pmatrix} = (1\ 3\ 2). \end{aligned}$$

We may do it from the left. In either case, we have to keep some information about intermediate states, which causes an extra burden on us. So, this strategy is not recommended for our purpose.

We return to our main theme of this section. From Theorem 5.11, we know that the symmetric group S_4 is generated by

$$\{(1\ 2), (1\ 3), (1\ 4)\}.$$

The following set is another system of generators:

$$\{\rho = (1\ 2\ 3\ 4), \sigma = (1\ 2\ 3), \tau = (1\ 2)\}.$$

Since $\{(1\ 2), (1\ 3)\}$ and $\{\sigma, \tau\}$ are both systems of generators of S_3, we can regard S_3 as a subgroup of S_4. We can characterize S_3 as a subgroup of S_4 that fixes the element 4. Then, we notice there are subgroups that fix 1, 2, or 3, respectively. These are conjugates of S_3.

We will find all normal subgroups of S_4. Let N be a normal subgroup of S_4.

(1) If N includes a cycle of length 4, then $N = S_4$. In fact, from Proposition 5.26, we know that N must contain all cycles of length 4. The cyclic group generated

by (1 2 3 4) is given by

$$\{e, (1\ 2\ 3\ 4), (1\ 3)(2\ 4), (1\ 4\ 3\ 2)\},$$

which must be contained in N. Then, all conjugates of (1 3)(2 4) must be contained in N. Take (1 2)(3 4) and compose as follows:

$$\begin{aligned} & (1\ 2\ 3\ 4) \quad (1\ 2)(3\ 4) \\ = & (1\ 4)(1\ 3)(1\ 2) \quad (1\ 2)(3\ 4) \\ = & (1\ 4)(1\ 3)(3\ 4) \\ = & (1\ 3). \end{aligned}$$

Then, all conjugates of (1 3), namely all transpositions must be contained in N, which constitute a system of generators of S_4.

(2) If N includes a cycle of length 3, then $N = A_4$. In fact, from Proposition 5.26, we know that N must contain all cycles of length 3.

Quiz 5.3 There are eight cycles of length 3 in S_4. Enumerate all.

Take two cycles of length 3 and multiply as follows:

$$(1\ 2\ 3)(1\ 2\ 4) = (1\ 3)(2\ 4).$$

So, N must contain all conjugates of (1 3)(2 4).

Quiz 5.4 There are three conjugates of (1 3)(2 4) including itself. Enumerate all.

The results of Quizes 5.3 and 5.4 provide jointly eleven permutations, which, if we add e, form the group of all even permutations, namely A_4. By construction, conjugates of any element is already contained in this subgroup, therefore A_4 is a normal subgroup of S_4. Note that we can also deduce A_4 is a normal subgroup from the fact that the size of A_4 is half of its parent's.

(3) If N includes a transposition $(a\ b)$ for some $1 \le a < b \le 4$, then N includes all transposition which generate the whole S_4. Therefore, we have $N = S_4$.

(4) If N includes the product of two disjoint transpositions $(a\ b)(c\ d)$, then taking all conjugates and adding e form a group

$$K_4 = \{e, (1\ 2)(3\ 4), (1\ 3)(2\ 4), (1\ 4)(2\ 3)\}.$$

A normal subgroup that includes some cycle of length 3 and the whole N must be the group of even permutations A_4. A normal subgroup that includes some transposition and the whole N must be S_4.

To sum up we have the following.

Proposition 5.29 *Normal subgroups of S_4 are K_4, A_4, and trivial subgroups only. Other subgroups are not normal.*

Exercises

[1] Express each of the following permutations in S_8 as a product of disjoint cycles, and find its order.

(1)

$$\begin{pmatrix} 1 & 2 & 3 & 4 & 5 & 6 & 7 & 8 \\ 5 & 4 & 6 & 2 & 7 & 8 & 1 & 3 \end{pmatrix}.$$

(2)

$$\begin{pmatrix} 1 & 2 & 3 & 4 & 5 & 6 & 7 & 8 \\ 5 & 7 & 2 & 6 & 1 & 3 & 8 & 4 \end{pmatrix}.$$

(3)

$$\begin{pmatrix} 1 & 2 & 3 & 4 & 5 & 6 & 7 & 8 \\ 3 & 2 & 5 & 7 & 1 & 6 & 8 & 4 \end{pmatrix}.$$

[2] Let us consider S_5.

(1) Enumerate all conjugates of $(1\ 2)$.
(2) Count the number of all conjugates of $(1\ 2)(3\ 4\ 5)$ including itself.
(3) Find a permutation $a \in S_5$ such that

$$a(4\ 5)(1\ 2\ 3)a^{-1} = (1\ 2)(3\ 4\ 5).$$

[3] We study the subgroups of S_4.

(1) Enumerate all subgroups of size 2. Classify them into conjugates.
(2) Enumerate all subgroups of size 3. Classify them into conjugates.
(3) Enumerate all subgroups of size 4. Classify them into conjugates.
(4) Enumerate all subgroups of size 6. Classify them into conjugates.
(5) Enumerate all subgroups of size 8. Classify them into conjugates.
(6) Enumerate all subgroups of size 12. Classify them into conjugates.

Chapter 6
Isomorphism Theorems for Groups

In the previous chapter, we calculated concrete examples of homomorphisms. Given a group homomorphism $\varphi : G \to H$, the factor group $G/\ker\varphi$ and $\operatorname{Im}\varphi$ seem in some sense identical. We will give this intuition a rigorous formulation in this chapter.

6.1 Isomorphism Theorems Help Us Find Homomorphisms

Since we have already seen many examples, we give a formal definition first.

Definition 6.1 If a group homomorphism $\varphi : G \to H$ is a bijection, it is called an *isomorphism*, and G and H are said *isomorphic*.

Remark 6.2 The inverse map $\varphi^{-1} : H \to G$ is also an isomorphism.

Theorem 6.3 (The First Isomorphism Theorem) *Let $\varphi : G \to H$ be a group homomorphism. We have the following:*

(1) $\ker\varphi$ *is a normal subgroup of* G.
(2) $\operatorname{Im}\varphi$ *is a subgroup of* H.
(3) $G/\ker\varphi$ *and* $\operatorname{Im}\varphi$ *are isomorphic.*

Remark 6.4 The isomorphism theorems will appear as the first, second, and third isomorphism theorems. We will see in this order.

Proof We have already discussed (1) and (2) in Sect. 4.2. In particular, (1) is Theorem 4.12. We prove (3) by constructing a bijection $\psi : G/\ker\varphi \to \operatorname{Im}\varphi$.

Any element of $G/\ker\varphi$ has an expression like $x(\ker\varphi)$. Set

$$\psi(x(\ker\varphi)) = \varphi(x) \in \operatorname{Im}\varphi.$$

S. Yukita, *First Lectures in Algebra*, Springer Asia Pacific Mathematics Series 10,
https://doi.org/10.1007/978-981-95-5457-7_6

We wonder if this "map" ψ is well-defined or not. To defeat this doubt, we have to show that the RHS does not depend on the choice of representatives. Let x' be any other representative on the same $\ker\varphi$-orbit, namely we have

$$x'(\ker\varphi) = x(\ker\varphi).$$

Multiplying both sides by x^{-1} from the left, we have

$$x^{-1}x'(\ker\varphi) = \ker\varphi,$$

which means there exists an element $h \in \ker\varphi$ such that

$$x^{-1}x' = h \in \ker\varphi.$$

Thus, we have $x' = xh$ and

$$\varphi(x'(\ker\varphi)) = xh(\ker\varphi) = x(\ker\varphi),$$

using $h(\ker\varphi) = \ker\varphi$. To sum up, we get

$$\psi(x'(\ker\varphi)) = \psi(x(\ker\varphi)).$$

(i) We show that ψ is injective. Let $x(\ker\varphi)$ and $y(\ker\varphi)$ be any two $\ker\varphi$-orbits satisfying

$$\psi(x(\ker\varphi)) = \psi(y(\ker\varphi)).$$

The definition of ψ gives us $\varphi(x) = \varphi(y)$. In Sect. 4.2, we established that $\ker\varphi$-orbits are fibers of φ. So, we have

$$x(\ker\varphi) = y(\ker\varphi).$$

This shows that ψ is injective.

(ii) We show that ψ is surjective. Given any element $y \in \operatorname{Im}\varphi$, there is an element $x \in G$ such that $\varphi(x) = y$. Then, we have

$$\psi(x(\ker\varphi)) = \varphi(x) = y,$$

which shows that ψ is surjective.

(iii) We show that ψ is a homomorphism. By definition, we have

$$\begin{aligned}
&\psi(x(\ker\varphi)\cdot y(\ker\varphi)) \\
=\ &\psi(xy(\ker\varphi))
\end{aligned}$$

$$= \varphi(xy) = \varphi(x)\varphi(y)$$
$$= \psi(x(\ker\varphi))\psi(y(\ker\varphi)).$$

We omit arguments on the identity and inverses.

We conclude that $\psi : G/\ker\varphi \to \operatorname{Im}\varphi$ is a group isomorphism. □

Remark 6.5 So far, we have used some design patterns for proofs.

- The definition of a group homomorphism using representatives must be well-defined. We use an alternative representative, recalculate, then verify the result does not change.
- *To show that a map is injective*, we first suppose that there are two elements that are sent to the same image. Then, we show that they are identical.
- *To show that a map is surjective*, for any given element of the codomain, we first seek a candidate element in the domain that is sent to the given element. Then, we verify that the image of the candidate coincides with the given element.

Be always conscious of design patterns. We can save time to find the first step.

Theorem 6.6 *Let $\varphi : G \to H$ be a group homomorphism. The map φ is injective if and only if* $\ker\varphi = \{e\}$.

Proof If φ is injective, all the fibers must be singleton sets. Especially, the fiber of e, namely $\ker\varphi$ must be $\{e\}$.

Conversely, if $\ker\varphi = \{e\}$, Theorem 4.11 tells us that all the fibers are singleton sets, which means φ is injective. □

Example 6.7 Let us find all the homomorphisms from the group $S_3 = \{e, \sigma, \sigma^2, \tau, \sigma\tau, \sigma^2\tau\}$ to $C_4 = \{e, a, a^2, a^3\}$ $(a^4 = e)$. Let $\varphi : S_3 \to C_4$ be a homomorphism.

We know that $\ker\varphi$ is a normal subgroup, which must be one of the following:

$$\{e\}, \quad \{e, \sigma, \sigma^2\}, \quad S_3.$$

(i) Suppose $\ker\varphi = \{e\}$. Then, φ is injective, which cannot be attained since $|S_3| > |C_4|$.
(ii) Suppose $\ker\varphi = \{e, \sigma, \sigma^2\}$. The quotient is $S_3/\ker\varphi \simeq C_2$, where C_2 is the cyclic group of size 2. Then, $\operatorname{Im}\varphi \simeq C_2$. The only subgroup of C_4 that is cyclic of size 2 is

$$\operatorname{Im}\varphi = \{e, a^2\}.$$

The homomorphism is determined as follows:

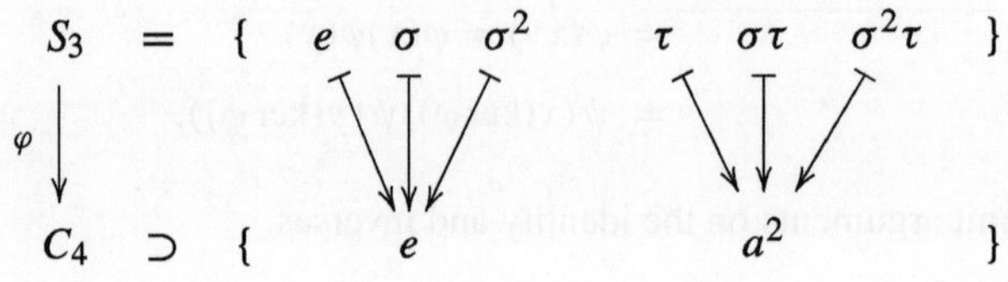

(iii) Suppose $\ker\varphi = S_3$. The homomorphism φ is a trivial one, which sends the whole S_3 to e.

We proceed to the second of the three isomorphism theorems.

Theorem 6.8 (The Second Isomorphism Theorem) *Let H be a subgroup of G, and N be a normal subgroup of G. Then, we have the following:*

(1) *$H \cap N$ is a normal subgroup of H.*
(2) *HN is a subgroup of G, and $HN = NH$ holds.*
(3) *N is a normal subgroup of HN.*
(4) *$H/(H \cap N) \simeq HN/N$.*

Let us begin with an example, setting aside the proof.

Example 6.9 Let $G = S_4$ and define $H = \{e, \rho, \rho^2, \rho^3\}$, where $\rho = (1\ 2\ 3\ 4)$. Then, H is explicitly given by

$$H = \{e, (1\ 2\ 3\ 4), (1\ 3)(2\ 4), (1\ 4\ 3\ 2)\},$$

which is not a normal subgroup of G. Let N be

$$N = K_4 = \{e, (1\ 2)(3\ 4), (1\ 3)(2\ 4), (1\ 4)(2\ 3)\},$$

which is a normal subgroup of G. See Proposition 5.29. We have the following:

$$H \cap N = \{e, (1\ 3)(2\ 4)\},$$
$$HN = \{e, \rho, \rho^2, \rho^3, (1\ 2)(3\ 4), (1\ 3), (1\ 4)(2\ 3), (2\ 4)\}.$$

(1) $H \cap N$ is a normal subgroup of H since $|H \cap N| = \frac{1}{2}|H|$. It is not a normal subgroup of G since some conjugate elements in G are missing in $H \cap N$. Note that $H \cap N$ has all conjugate elements in H. **What is conjugate to what depends on the parent (ambient) group.**
(2) HN is a subgroup of G, and $HN = NH$ holds. It is not a normal subgroup of G since some conjugate elements in G are missing in HN.
(3) N is a normal subgroup of HN since $|N| = \frac{1}{2}|HN|$.
(4) $H/(H \cap N) \simeq HN/N$. See Fig. 6.1 to confirm

$$H/(H \cap N) = \{\{e, \rho^2\}, \{\rho, \rho^3\}\},$$

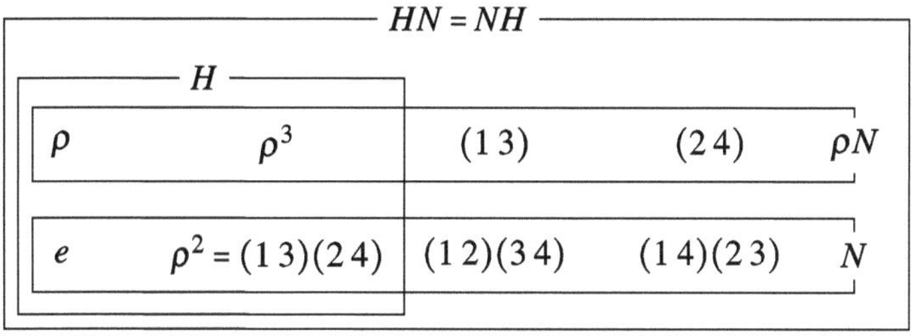

Fig. 6.1 $H/(H \cap N) \simeq HN/N$

$$HN/N = \{N, \rho N\}.$$

The isomorphism is given by the following:

$$\begin{array}{ccc} H/(H \cap N) & \longrightarrow & HN/N \\ H \cap N & \mapsto & N \\ \rho(H \cap N) & \mapsto & \rho N \end{array}$$

Quiz 6.1 Calculate $H \cap N$ and HN in Example 6.9.

Proof (The Second Isomorphism Theorem)

(1) Since $xH = Hx = H$ for all $x \in H$, we have

$$x(H \cap N) = xH \cap xN = H \cap xN,$$
$$(H \cap N)x = Hx \cap Nx = H \cap Nx.$$

Since N is a normal subgroup of G, we have $xN = Nx$. Thus, we have

$$x(H \cap N) = (H \cap N)x$$

for all $x \in H$, which means that $H \cap N$ is a normal subgroup of H.

(2) Take two arbitrary elements h_1n_1 and h_2n_2 from $HN = \{hn \mid h \in H, n \in N\}$. We want to see if their product $(h_1n_1)(h_2n_2)$ is in HN. Since N is a **normal subgroup** of G, we have $Nh_2 = h_2N$. Therefore, there is an element $n_3 \in N$ such that $n_1h_2 = h_2n_3$. Using this, we have

$$(h_1n_1)(h_2n_2) = h_1(n_1h_2)n_2 = h_1(h_2n_3)n_2 = (h_1h_2)(n_3n_2) \in HN,$$

which shows that HN is closed under multiplication.

Next, we show that HN is closed under inversion. Let $h^{-1}n_?$ be a candidate for the inverse of $hn \in HN$. Since N is a **normal subgroup** of G, we have

$Nh^{-1} = h^{-1}N$. Therefore, there is an element $n_0 \in N$ such that $nh^{-1} = h^{-1}n_0$. Thus, we have

$$\begin{aligned}&(hn)(h^{-1}n_?)\\&= h(nh^{-1})n_?\\&= h(h^{-1}n_0)n_?\\&= (hh^{-1})(n_0n_?)\\&= n_0n_?.\end{aligned}$$

If we set $n_? = n_0^{-1}$, then we have

$$(hn)^{-1} = h^{-1}n_0^{-1},$$

which shows that HN is closed under inversion.

We conclude that HN is a subgroup of G.

Next, we prove $HN = NH$. Let $hn \in HN$. Since N is a normal subgroup of G, there is an element $n' \in N$ such that $hn = n'h$. We have $hn = n'h \in NH$. Therefore, $HN \subset NH$. The same argument gives also $NH \subset HN$. Thus, we have $HN = NH$.

(3) It is clear that N is a subgroup of HN. Since N is a normal subgroup of G, for all $x \in G$ we have $Nx = xN$, therefore for all $x \in HN \subset G$ we have $Nx = xN$. Thus, N is a normal subgroup of HN.

(4) We first construct a homomorphism $\psi : H/(H \cap N) \to HN/N$ by taking the correspondence as

$$\psi : h(H \cap N) \mapsto hN.$$

To see that this is well-defined, we replace h by an alternative representative $h_1 = hn$ for some $n \in H \cap N$. We have

$$h_1N = hnN = hN,$$

which shows the correspondence does not depend on the choice of representatives.

(i) We show that ψ is a homomorphism. Take any elements $h_1(H \cap N)$ and $h_2(H \cap N)$ from $H/(H \cap N)$. Using the definition of ψ, we have

$$\begin{aligned}&\psi(h_1(H \cap N) \cdot h_2(H \cap N))\\&= \psi(h_1h_2(H \cap N))\\&= h_1h_2N = h_1N \cdot h_2N\end{aligned}$$

$$= \psi(h_1(H \cap N)) \cdot \psi(h_2(H \cap N)),$$

where we use the facts that N is a normal subgroup of HN and that $H \cap N$ is a normal subgroup of H. Arguments on the identity element and inverses go in the same way.

(ii) We show that ψ is injective. Theorem 6.6 tells us that it is sufficient to show $\ker \psi = \{e(H \cap N)\}$. If $h(H \cap N) \in \ker \psi$, then we have

$$\psi(h(H \cap N)) = hN = N,$$

which means $h \in N$. Since $h \in H$, we have $h \in H \cap N$. Then,

$$h(H \cap N) = e(H \cap N).$$

(iii) We show that ψ is surjective. Take an arbitrary element from HN/N, say hnN with $h \in H$ and $n \in N$. We immediately see that $hnN = hN$. We have

$$\psi(h(H \cap N)) = hN,$$

which shows ψ is surjective.

From (i), (ii), and (iii), we conclude that $\psi : H/(H \cap N) \to HN/N$ is an isomorphism.

□

Remark 6.10 In the second isomorphism theorem, the normality of the subgroup N is essential. Let us consider the following example. Let $B_1 = \{e, \tau\}$ and $B_2 = \{e, \sigma\tau\}$ be subgroups of S_3. We have the following:

$$B_1 B_2 = \{e, \tau, \sigma\tau, \sigma^2\},$$
$$B_2 B_1 = \{e, \tau, \sigma\tau, \sigma\}.$$

We see that $B_1 B_2 \neq B_2 B_1$. Neither of them is a subgroup of S_3. In the first place, there is no subgroup of size 4 in S_3 by Lagrange's theorem 3.17.

The following corollary is a specialization of the second isomorphism theorem.

Corollary 6.11 *Let H be a subgroup of G, and N be a normal subgroup of G. If $H \supset N$, then we have:*

(1) *N is a normal subgroup of H.*
(2) *H/N is a subgroup of G/N.*

Remark 6.12 Since N is a normal subgroup of G, the left and right N-orbits starting the same element are identical. We assumed $N \subset H$, which means that

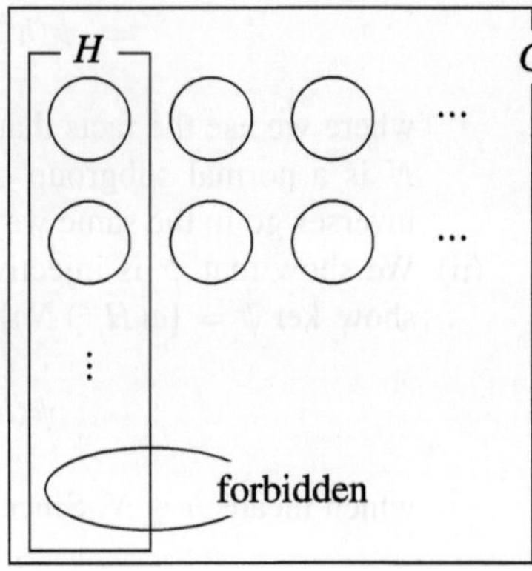

Fig. 6.2 N-orbits

any N-orbit does not cross the border of H and $G - H$. In other words, if an N-orbit starts at an element of H, it remains in H. If an N-orbit starts at an element outside of H, it remains outside of H. See Fig. 6.2.

Example 6.13 Let $\mathbb{Z}$ be the set of all integers. We consider $\mathbb{Z}$ as a group under addition. Since the group $\mathbb{Z}$ is abelian, all subgroups are normal. Let $2\mathbb{Z}$, $3\mathbb{Z}$, and $6\mathbb{Z}$ be the subgroups of $\mathbb{Z}$ consisting of the multiples of 2, 3, and 6, respectively. We have

$$\mathbb{Z}/6\mathbb{Z} = \{6\mathbb{Z}, 6\mathbb{Z} + 1, \ldots, 6\mathbb{Z} + 5\}.$$

We write $6\mathbb{Z} + k = \bar{k}$. Then, we have the following:

$$2\mathbb{Z}/6\mathbb{Z} = \{\bar{0}, \bar{2}, \bar{4}\},$$
$$3\mathbb{Z}/6\mathbb{Z} = \{\bar{0}, \bar{3}\}.$$

6.2 Subgroups in Quotients

We study subgroups in factor groups. Let N be a normal subgroup of G. We consider the following two sets of subgroups:

$$X = \{H \mid H \text{ is a subgroup of } G,\ H \supset N\},$$
$$Y = \{J \mid J \text{ is a subgroup of } G/N\}.$$

Let of H be any element of X, namely a subgroup H of G that contains N. Then, N is a normal subgroup of H.

Quiz 6.2 Prove that N is a normal subgroup of H above.

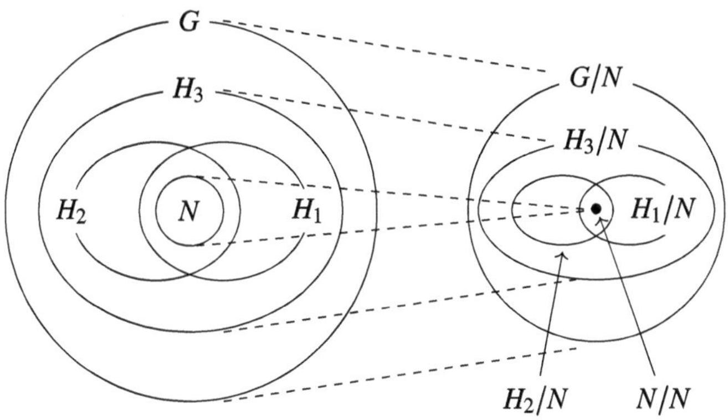

Fig. 6.3 Canonical projection map

We define $\varphi : X \to Y$ by

$$\varphi : H \mapsto H/N.$$

See Fig. 6.3 to get a feel.

Theorem 6.14 (The Third Isomorphism Theorem) *The map $\varphi : X \to Y$ is a bijection which preserves the partial order by inclusion. Namely, it has the following properties:*

(1) *φ is a bijection.*
(2) *If $H_1 \subset H_2$, then $\varphi(H_1) \subset \varphi(H_2)$.*
(3) *$\varphi(H_1 \cap H_2) = \varphi(H_1) \cap \varphi(H_2)$.*

Proof We first prove that φ is bijective. Then, we prove the partial order by inclusion is preserved.

(1i) We show that φ is injective. Let $H_1, H_2 \in X$ and $H_1/N = H_2/N$. Since both H_1 and H_2 are disjoint unions of N-orbits, we have $H_1 = H_2$.
(1s) We show that φ is surjective. Let J be any subgroup of G/N. If we set

$$H = \{x \in G \mid xN \in J\},$$

then we have

$$N \subset H \subset G.$$

We show that H is a subgroup of G. Take arbitrary elements $x, y \in H$. We have

$$xN \in J, \quad \text{and} \quad yN \in J.$$

Since J is a subgroup of G/N, we have

$$xyN = (xN)(yN) \in J.$$

This shows that $xy \in H$. We have also

$$x^{-1}N = (xN)^{-1} \in J,$$

which means $x^{-1} \in H$. We see that H is a subgroup of G that contains N. Thus, we have

$$\varphi(H) = J.$$

(2) We show that ψ preserves the inclusion relation. If $N \subset H_1 \subset H_2$, then

$$H_1/N \subset H_2/N,$$

which means

$$\varphi(H_1) \subset \varphi(H_2).$$

(3) Since N is a normal subgroup of G, it is also a normal subgroup of H_1, H_2, and $H_1 \cap H_2$, each of which is a disjoint union of N-orbits. Thus, we have

$$(H_1 \cap H_2)/N = (H_1/N) \cap (H_2/N),$$

which means

$$\varphi(H_1 \cap H_2) = \varphi(H_1) \cap \varphi(H_2).$$

□

Example 6.15 Let $G = \mathbb{Z}$ be the group of integers under addition. Since G is abelian, all subgroups are normal. Let $N = 12\mathbb{Z}$ be a subgroup of $\mathbb{Z}$ consisting of all multiples of 12. Since G/N is abelian, all subgroups are normal. Let us write

$$G/N = \{\bar{0}, \bar{1}, \bar{2}, \ldots, \overline{11}\}.$$

Subgroups of G that contains N are enumerated as

$$X = \{12\mathbb{Z}, 6\mathbb{Z}, 4\mathbb{Z}, 3\mathbb{Z}, 2\mathbb{Z}, \mathbb{Z}\}.$$

Subgroups of G/N are enumerated as

$$Y = \{\{\bar{0}\}, \{\bar{0}, \bar{6}\}, \{\bar{0}, \bar{4}, \bar{8}\}, \{\bar{0}, \bar{3}, \bar{6}, \bar{9}\}, \{\bar{0}, \bar{2}, \ldots, \overline{10}\}, G/N\}.$$

The map $\varphi : X \to Y$ is given as follows:

$$\begin{aligned} \varphi : \quad X &\longrightarrow Y \\ 12\mathbb{Z} &\mapsto \{\bar{0}\} \\ 6\mathbb{Z} &\mapsto \{\bar{0}, \bar{6}\} \\ 4\mathbb{Z} &\mapsto \{\bar{0}, \bar{4}, \bar{8}\} \\ 3\mathbb{Z} &\mapsto \{\bar{0}, \bar{3}, \bar{6}, \bar{9}\} \\ 2\mathbb{Z} &\mapsto \{\bar{0}, \bar{2}, \ldots, \overline{10}\} \\ \mathbb{Z} &\mapsto G/N \end{aligned}$$

Exercises

[1] Find all homomorphisms between the given pair of groups.

(1) $S_3 \to C_6 = \{e, a, \ldots, a^5\}$.
(2) $D_4 \to C_4 = \{e, c, c^2, c^3\}$.
(3) $D_5 \to C_{10} = \{e, c, \ldots, c^9\}$.
(4) $Q_8 \to C_4 = \{e, c, c^2, c^3\}$.

[2] Take G in the second isomorphism theorem as

$$D_4 = \{e, a, a^2, a^3, b, ab, a^2b, a^3b\},$$

where $a^i b = ba^{4-i}$. Let N be a normal subgroup of D_4 given by

$$N = \{e, a, a^2, a^3\}.$$

Let H be a subgroup of D_4 given by

$$H = \{e, a^2, b, a^2b\}.$$

Applying the second isomorphism theorem to these groups $G = D_4$, N, and H, find a concrete expression for

$$H/(H \cap N) \to HN/N.$$

[3] Consider a normal subgroup N of S_4 given by

$$N = \{e, (1\ 2)(3\ 4), (1\ 3)(2\ 4), (1\ 4)(2\ 3)\}.$$

(1) Write down S_4/N.

(2) Prove $S_4/N \simeq S_3$ by showing that we can pick out representatives from

$$S_3 = \{e, (1\ 2\ 3), (1\ 3\ 2), (1\ 2), (1\ 3), (2\ 3)\}.$$

[4] Consider the factor group S_4/N and the correspondence between the sets X and Y given as follows:

$$X = \{H \mid H \text{ is a subgroup of } S_4,\ H \supset N\},$$
$$Y = \{J \mid J \text{ is a subgroup of } S_4/N\}.$$

(1) Write out all elements of the set X.
(2) Write out all elements of the set Y.
(3) Give a bijection $\varphi : X \to Y$ that preserves the inclusion relation and the intersection.

Chapter 7
Products of Groups

Viewing a complex object as a composite of simple objects is a common practice in various disciplines. The prime decomposition of natural numbers is the most prominent example. We will see that similar things can be done in group theory.

7.1 Products

We introduce the notion of products of groups. We utilize the notion in studying the structures of finite cyclic groups.

Let us begin with examples.

Example 7.1 Compare the following groups:

$$\begin{aligned} C_4 &= \{e, a, a^2, a^3\}, \text{ where } \quad a^4 = e, \\ K_4 &= \{e, x, y, xy\}, \text{ where } \quad x^2 = y^2 = e, \ \ yx = xy. \end{aligned}$$

Both are abelian groups of size 4, but they are clearly not isomorphic. The group K_4 is called the *Klein four group*.

Quiz 7.1 Write out multiplication tables of C_4 and K_4. Give evidence that they are not isomorphic.

The group C_4 is a cyclic group. Whereas K_4 is not cyclic. We want to get a clear view of this difference, using the notion of products.

Definition 7.2 Let G and H be groups. We consider the cartesian product

$$G \times H = \{(x, y) \mid x \in G, y \in H\}$$

endowed with a binary operation given by

S. Yukita, *First Lectures in Algebra*, Springer Asia Pacific Mathematics Series 10,
https://doi.org/10.1007/978-981-95-5457-7_7

$$(x_1, y_1)(x_2, y_2) = (x_1x_2, y_1y_2) \in G \times H$$

for any $(x_1, y_1), (x_2, y_2) \in G \times H$. We can easily see that $G \times H$ is a group under this operation. The identity element is (e, e). The inverse of (x, y) is (x^{-1}, y^{-1}). The group $G \times H$ is called the *product of* G and H.

Example 7.3 Let $C_2 = \{e, z\}$ be a cyclic group of size 2. Consider the product

$$C_2 \times C_2 = \{(e, e), (z, e), (e, z), (z, z)\}.$$

We can construct an isomorphism between K_4 (in Example 7.1) and $C_2 \times C_2$ as follows:

$$\begin{aligned} K_4 &\to C_2 \times C_2 \\ e &\mapsto (e, e) \\ x &\mapsto (z, e) \\ y &\mapsto (e, z) \\ xy &\mapsto (z, z) \end{aligned}$$

Thus, we have $K_4 \simeq C_2 \times C_2$.

Example 7.4 Consider the group

$$C_6 = \{e, a, \ldots, a^5\}, \quad a^6 = e$$

and its subgroups

$$C_2 = \{e, a^3\} \quad \text{and} \quad C_3 = \{e, a^2, a^4\}.$$

We can see that (a^3, a^2) is a generator of the group $C_2 \times C_3$ as follows:

$$\begin{aligned} (a^3, a^2)^2 &= (a^6, a^4) = (e, a^4), \\ (a^3, a^2)^3 &= (a^3, e), \\ (a^3, a^2)^4 &= (e, a^2), \\ (a^3, a^2)^5 &= (a^3, a^4), \\ (a^3, a^2)^6 &= (e, e). \end{aligned}$$

This sequence of powers shows that the order of (a^3, a^2) is 6. Thus, the group $C_2 \times C_3$ is a cyclic group. We can construct an isomorphism $C_2 \times C_3 \to C_6$ as follows:

$$\begin{array}{ccc} C_2 \times C_3 & \longrightarrow & C_6 \\ (e, e) & \mapsto & e \\ (a^3, a^2) & \mapsto & a^5 \\ \vdots & & \vdots \\ (a^3, a^2)^5 & \mapsto & a \end{array}$$

Alternatively, we can set $(a^3, a^2) \mapsto a$ and complete the remaining correspondences. Either choice gives $C_6 \simeq C_2 \times C_3$.

7.2 Structures of Finite Cyclic Groups

We begin by collecting simple facts.

Proposition 7.5 *Two cyclic groups of the same size are isomorphic. This holds also when the size is infinity.*

Proof Let A and B be cyclic groups of size n. Let a and b be generators of A and B, respectively. The map $\varphi : A \to B$

$$\varphi : a^k \mapsto b^k$$

is an isomorphism. □

Proposition 7.6 *Let m and n be natural numbers that are mutually prime. Then, we have*

$$C_m \times C_n \simeq C_{mn}.$$

Proof Let $a \in C_m$ and $b \in C_n$ be generators of C_m and C_n, respectively. The order of $(a, b) \in C_m \times C_n$ is mn. Since the size of the product $C_m \times C_n$ is mn, the element (a, b) is a generator, therefore the product is a cyclic group. By Proposition 7.5, we conclude

$$C_m \times C_n \simeq C_{mn}.$$

□

Theorem 7.7 *Let a natural number n have the prime decomposition*

$$n = 2^{t_1} \times 3^{t_2} \times \cdots \times p_k^{t_k}.$$

We have

$$C_n \simeq C_{2^{t_1}} \times C_{3^{t_2}} \times \cdots \times C_{p_k{}^{t_k}},$$

where p_i is the i-th smallest prime, for example $p_1 = 2$, $p_2 = 3$, $p_3 = 5$.

Proof Use Proposition 7.6 repeatedly. □

Quiz 7.2 Explain why $C_{12} \not\simeq C_6 \times C_2$.

So far, the discussion of this chapter is limited to finite cyclic groups. The next steps of study in this direction might be:

- structures of finite abelian groups,
- structures of finitely generated abelian groups.

In the remainder of this chapter, we will see the structure theorem of finitely generated abelian groups, skipping proofs.

Definition 7.8 Let G be an abelian group. If a finite subset $\{a_1, a_2, a_3, \cdots, a_s\} \subset G$ generates G, namely for any $x \in G$ there are integers $n_1, n_2, \ldots, n_s$ such that

$$x = a_1^{n_1} a_2^{n_2} \cdots a_s^{n_s},$$

then the group G is called a *finitely generated abelian group*. The set $\{a_1, a_2, a_3, \cdots, a_s\}$ is called a *system of generators*.

Example 7.9 Easy to see facts are listed below.

(1) Finite abelian groups are all finitely generated.
(2) An infinite cyclic group C_∞ is not a finite group but clearly a finitely generated abelian group.
(3) The group $\mathbb{Q}$ under addition is not finitely generated. Assume a finite set $\{a_1, a_2, a_3, \cdots, a_s\} \subset \mathbb{Q}$ generates $\mathbb{Q}$. Then, for any $x \in \mathbb{Q}$ there should be integers $n_1, n_2, \cdots, n_s$ such that

$$x = n_1 a_1 + n_2 a_2 + \cdots + n_s a_s.$$

Let m be the least common multiple of the denominators of $a_1, a_2, a_3, \cdots, a_s$. Then, we have

$$\mathbb{Z}a_1 + \mathbb{Z}a_2 + \cdots + \mathbb{Z}a_s \subset \mathbb{Z}\frac{1}{m} \subsetneq \mathbb{Q},$$

which clearly shows that $a_1, a_2, a_3, \cdots, a_s$ cannot generate $\mathbb{Q}$.
(4) The group $\mathbb{R}$ under addition is not finitely generated, which is clear from (3).

The structure theorem is stated in one of the two formulations.

Theorem 7.10 (Primary Decomposition Version) *Given any finitely generated abelian group G, there are natural numbers $n, q_1, q_2, \cdots, q_t$ such that*

$$G \simeq C_{\infty}^{n} \times C_{q_1} \times C_{q_2} \times \cdots \times C_{q_t}.$$

Here,

- *The natural number n is uniquely determined by G.*
- *Every q_i is a power of some prime p_i. We allow cases in which $p_i = p_j$ for $i \neq j$.*
- *$q_1, q_2, \cdots, q_t$ are uniquely determined by G up to ordering.*

It is intuitively trivial, but proving it takes pages.

Theorem 7.11 (Invariant Factors Version) *Given any finitely generated abelian group G, there are natural numbers $n, k_1, k_2, \cdots, k_t$ such that*

$$G \simeq C_{\infty}^{n} \times C_{k_1} \times C_{k_2} \times \cdots \times C_{k_t},$$

where

$$k_1 \mid k_2 \mid k_3 \mid \cdots \mid k_{t-1} \mid k_t.$$

Remark 7.12 The binary relation $a \mid b$ means that a divides b. The natural numbers $k_1, k_2, \cdots, k_t$ are called *invariant factors*.

Example 7.13 Let $G = C_2 \times C_4 \times C_6$. The primary decomposition is given by

$$C_2^2 \times C_3 \times C_4.$$

The invariant factor decomposition is given by

$$C_2 \times C_2 \times C_{12}.$$

Exercises

[1] Show that $C_{15} \simeq C_3 \times C_5$.
[2] We have $C_9 \not\simeq C_3 \times C_3$. Show that the RHS is not a cyclic group.
[3] Show that $C_{12} \simeq C_3 \times C_4$.
[4] Consider the group $S_3 = \{e, \sigma, \sigma^2, \tau, \sigma\tau, \sigma^2\tau\}$ and its subgroup

$$A = \{e, \sigma, \sigma^2\}, \quad B = \{e, \tau\}.$$

Show that $S_3 \not\simeq A \times B$.

Chapter 8
Rings and Fields

A group is an algebraic system with one binary operation. A ring or field is an algebraic system with two binary operations, multiplication and addition, with some constraints between them. A module over a ring is a generalization of a vector space over a field.

8.1 Rings and Fields

The notions of ring and field are abstracted from familiar algebraic systems with two binary operations, addition and multiplication, with distributive laws, such as integers, rationals, reals, and complexes.

Definition 8.1 A set A endowed with two binary operations $+$ and $\times$ is called a *ring* if it satisfies the following conditions:

(1) A is an abelian group under the binary operation $+$ called *addition*. Its neutral element is called the *zero element* and denoted often by 0.
(2) The binary operation $\times$ called *multiplication* is associative and has a neutral element called the *unity* denoted often by 1.
(3) For any $x, y, z \in A$ we have

$$(x + y) \times z = x \times z + y \times z,$$
$$z \times (x + y) = z \times x + z \times y.$$

We call this condition the *distributive law*.

If the binary operation $\times$ is commutative, then the ring is called a *commutative ring*.

Remark 8.2 We often omit the symbol $\times$ in the expressions.

S. Yukita, *First Lectures in Algebra*, Springer Asia Pacific Mathematics Series 10,
https://doi.org/10.1007/978-981-95-5457-7_8

Example 8.3 Some rings are listed below including those we are familiar with.

$\mathbb{Z}$	The ring of integers
$\mathbb{Q}$	The ring of rationals
$\mathbb{R}$	The ring of reals
$\mathbb{C}$	The ring of complexes
$M_n(\mathbb{C})$	The ring of $n \times n$ complex matrices
$\mathbb{Z}[X]$	The polynomial ring of X over $\mathbb{Z}$
$\mathbb{Q}[X]$	The polynomial ring of X over $\mathbb{Q}$
$\mathbb{C}[X]$	The polynomial ring of X over $\mathbb{C}$
$\mathbb{C}[X, Y]$	The polynomial ring of X and Y over $\mathbb{C}$
$\mathbb{Z}[\sqrt{-1}]$	$\{f(\sqrt{-1}) \mid f(X) \in \mathbb{Z}[X]\}$ $= \{m + n\sqrt{-1} \mid m, n \in \mathbb{Z}\}$ since $\sqrt{-1}^2 = -1$
$\mathbb{Q}[\sqrt{2}]$	$\{f(\sqrt{2}) \mid f(X) \in \mathbb{Q}[X]\}$ $= \{p + q\sqrt{2} \mid p, q \in \mathbb{Q}\}$ since $\sqrt{2}^2 = 2$.

Let A be one of $\mathbb{Z}$, $\mathbb{Q}$, $\mathbb{R}$, and $\mathbb{C}$. When we say $f(X)$ is a polynomial over A, it means a polynomial of X with coefficients in A.
The ring $M_n(\mathbb{C})$ is a non-commutative ring when $n \geq 2$.

Quiz 8.1 Verify that each set in the above list is closed under addition and multiplication. Find the unity for each.

Remark 8.4 Constructing a polynomial ring over some ring A is a means of *creating a new ring out of a known one*. Take the ring of rational numbers $\mathbb{Q}$. The polynomial ring of X over $\mathbb{Q}$ is given by

$$\mathbb{Q}[X] = \{a_0X^n + a_1X^{n-1} + \cdots + a_{n-1}X + a_n \mid a_k \in \mathbb{Q}\},$$

with usual addition and multiplication in high-school mathematics. The same constructions are possible for $\mathbb{Z}$, $\mathbb{R}$, $\mathbb{C}$, or any other rings. The symbol X is usually called an *indeterminate*.

Combining substitution and polynomial construction provides another means of creating a new ring out of known ones. Consider the ring

$$\mathbb{Z}[\sqrt{-1}] = \{f(\sqrt{-1}) \mid f(X) \in \mathbb{Z}[X]\}$$

in Example 8.3. We can express this in an alternative form like

$$\mathbb{Z}[\sqrt{-1}] = \{a_0\sqrt{-1}^n + a_1\sqrt{-1}^{n-1} + \cdots + a_{n-1}\sqrt{-1} + a_n \mid a_k \in \mathbb{Z}\}.$$

Since $\sqrt{-1}^2 = -1$, each term of an even power of $\sqrt{-1}$ is reduced to an integer, and each term of an odd power of $\sqrt{-1}$ is reduced to the product of an integer and

$\sqrt{-1}$. Therefore, we have

$$\mathbb{Z}[\sqrt{-1}] = \{m + n\sqrt{-1} \mid m, n \in \mathbb{Z}\}.$$

Consider the ring

$$\{f(\sqrt{2}) \mid f(X) \in \mathbb{Q}[X]\}.$$

We can express this in an alternative form like

$$\mathbb{Q}[\sqrt{2}] = \{a_0\sqrt{2}^n + a_1\sqrt{2}^{n-1} + \cdots + a_{n-1}\sqrt{2} + a_n \mid a_k \in \mathbb{Q}\}.$$

Since $\sqrt{2}^2 = 2$, each term of an even power of $\sqrt{2}$ is reduced to a rational number, and each term of an odd power of $\sqrt{2}$ is reduced to the product of a rational number and $\sqrt{2}$. Therefore, we have

$$\mathbb{Q}[\sqrt{2}] = \{p + q\sqrt{2} \mid p, q \in \mathbb{Q}\}.$$

Remark 8.5 Like the ring $M_n(\mathbb{C})$ in the list above, we can add $M_n(\mathbb{Z})$, $M_n(\mathbb{Q})$, $M_n(\mathbb{R})$, and $M_n(A)$ in general for any ring A.

Example 8.6 Let n be a natural number. Let us consider modular arithmetic modulo n. We introduce addition $+$ and multiplication $\times$ to the set $\mathbb{Z}_n = \{0, 1, \ldots, n-1\}$ by wrapping around when the results exceed n. The addition and multiplication tables of $\mathbb{Z}_6$ are given below.

+	0	1	2	3	4	5
0	0	1	2	3	4	5
1	1	2	3	4	5	0
2	2	3	4	5	0	1
3	3	4	5	0	1	2
4	4	5	0	1	2	3
5	5	0	1	2	3	4

×	1	2	3	4	5
1	1	2	3	4	5
2	2	4	0	2	4
3	3	0	3	0	3
4	4	2	0	4	2
5	5	4	3	2	1

We usually exclude 0 from the title row and column of the multiplication table of a ring to save space. We immediately notice that the group of addition is a cyclic group. We believe now the distributive law holds, which will be discussed systematically later. So, $\mathbb{Z}_6$ is a ring.

The tables of $\mathbb{Z}_5$ are given below.

The group of addition is a cyclic group. The set $\mathbb{Z}_5 - \{0\}$ is also a cyclic group under multiplication. We can take 2 or 3 as a generator. The situation around multiplication is quite different from $\mathbb{Z}_6$. This gives rise to the notion of a field, coming shortly.

+	0	1	2	3	4
0	0	1	2	3	4
1	1	2	3	4	0
2	2	3	4	0	1
3	3	4	0	1	2
4	4	0	1	2	3

×	1	2	3	4
1	1	2	3	4
2	2	4	1	3
3	3	1	4	2
4	4	3	2	1

Quiz 8.2 Verify that the set $\mathbb{Z}_5 - \{0\}$ is closed under multiplication, and that it is a cyclic group.

The notion of subgroups were introduced in group theory. Likewise, we introduce the notion of subrings in ring theory.

Definition 8.7 Let A be a ring and B be a subgroup of the addition group of A. The set B is called a *subring* of A if it satisfies the following conditions:

(1) For any $x, y \in B$ we have $xy \in B$. Namely B is closed under multiplication.
(2) $1 \in B$.

Example 8.8 The following sequence of inclusions is the subring relation transitively:

$$\mathbb{Z} \subset \mathbb{Q} \subset \mathbb{R} \subset \mathbb{C}.$$

The ring $\mathbb{C}$ is a subring of $\mathbb{C}[X]$ if we regard a complex number as a polynomial consisting only of a constant term. The ring $\mathbb{C}[X]$ is a subring of $\mathbb{C}[X, Y]$ if we regard a polynomial of X as a polynomial of X and Y lacking terms involving Y. The ring $\mathbb{Z}[X]$ is a subring of $\mathbb{Q}[X]$.

Example 8.9 The subgroup $n\mathbb{Z}$ ($n \geq 2$ is a natural number) of the group of addition is not a subring since it lacks the unity, which violates condition (2) in Definition 8.7.

Example 8.10 Let A and B be rings. When we say that B is a subring of A, condition (2) in Definition 8.7 excludes the following case. We define the rings A and B by

$$A = \{(x, y) \mid x, y \in \mathbb{Z}\}, \quad B = \{(x, 0) \mid x \in \mathbb{Z}\},$$

with addition and multiplication defined by

$$(x_1, y_1) + (x_2, y_2) = (x_1 + x_2, y_1 + y_2), \quad (x_1, y_1) \times (x_2, y_2) = (x_1x_2, y_1y_2).$$

The zero element A is $(0, 0)$, and the unity $(1, 1)$. If we restrict these componentwise operations to B, we get a ring B whose zero element is $(0, 0)$ and unity $(1, 0)$. Notice that the unity $(1, 1)$ of A is not an element of B. Such a situation will be treated with the notion of modules in later chapters.

Next, we consider the inverse operation of multiplication in rings.

Definition 8.11 Let A be a ring. An element $x \in A$ is called a *unit* if is has a multiplicative inverse x^{-1}. The set of all units in A is denoted by $A^\times$. A ring A is called a *division ring* if all the non-zero elements are units. If A is a division ring, then we have

$$A^\times = A - \{0\} = \{x \in A \mid x \neq 0\}.$$

If the multiplication $\times$ is commutative, then the division ring A is called a *field*. If the multiplication $\times$ is non-commutative, then the division ring A is called a *skew field*.

Remark 8.12 We have

$$\{\text{division rings}\} = \{\text{fields}\} \sqcup \{\text{skew fields}\}.$$

In this book we use this terminology and convention. However, in some books we may encounter the following terminology:

$$\{\text{fields}\} = \{\text{commutative fields}\} \sqcup \{\text{skew fields}\}, \quad \text{or}$$
$$\{\text{skew fields}\} = \{\text{commutative fields}\} \sqcup \{\text{non-commutative fields}\}.$$

The reader should take note of the diversity in terminology for future advanced study.

Proposition 8.13 *Let A be a ring. The set of units $A^\times$ is a group under multiplication.*

Proof Associativity and the existence of the unity, the neutral element of multiplication, are guaranteed by the definition of rings. Let $x, y \in A^\times$. Clearly, both x^{-1} and y^{-1} are units. The element xy is also a unit since $(xy)^{-1} = y^{-1}x^{-1}$. □

Example 8.14 The rings $\mathbb{Q}$, $\mathbb{R}$, $\mathbb{C}$, and $\mathbb{Z}_5$ are fields. The rings $\mathbb{Z}$, $M_n(\mathbb{C})$ for $n \geq 2$, $\mathbb{Q}[X]$, $\mathbb{C}[X]$, $\mathbb{Z}_6$ are not division rings.

Example 8.15 We have seen matrix rings $M_n(k)$ ($n \geq 2$) as examples of non-commutative rings. We add a new example, a skew field (or non-commutative division ring) of *quaternions*. We consider a 4-dimensional real vector space $\mathbb{H}$ with a basis $\{1, \mathbf{i}, \mathbf{j}, \mathbf{k}\}$. Namely, we have

$$\mathbb{H} = \{u + x\mathbf{i} + y\mathbf{j} + z\mathbf{k} \mid u, x, y, z \in \mathbb{R}\}.$$

We define multiplication among the members of the basis as follows:

$$\mathbf{ij} = -\mathbf{ji} = \mathbf{k}, \quad \mathbf{jk} = -\mathbf{kj} = \mathbf{i}, \quad \mathbf{ki} = -\mathbf{ik} = \mathbf{j}, \quad \mathbf{i}^2 = \mathbf{j}^2 = \mathbf{k}^2 = -1.$$

We extend this multiplication to the whole $\mathbb{H}$ distributively, namely we have

$$\begin{aligned}&(u + x\mathbf{i} + y\mathbf{j} + z\mathbf{k})(v + p\mathbf{i} + q\mathbf{j} + r\mathbf{k})\\&\stackrel{\text{def}}{=} uv - xp - yq - zr\\&\quad + (up + xv + yr - zq)\mathbf{i} + (uq + yv + zp - xr)\mathbf{j} + (ur + zv + xq - yp)\mathbf{k}.\end{aligned}$$

The unity is 1. For any $u + x\mathbf{i} + y\mathbf{j} + z\mathbf{k} \neq 0$, we want a concrete expression for $(u+x\mathbf{i}+y\mathbf{j}+z\mathbf{k})^{-1}$. We have to solve the system of linear equations with unknowns v, p, q, r below:

$$\begin{pmatrix} u & -x & -y & -z \\ x & u & -z & y \\ y & z & u & -x \\ z & -y & x & u \end{pmatrix} \begin{pmatrix} v \\ p \\ q \\ r \end{pmatrix} = \begin{pmatrix} 1 \\ 0 \\ 0 \\ 0 \end{pmatrix}. \tag{8.1}$$

The result is as follows:

$$(u + x\mathbf{i} + y\mathbf{j} + z\mathbf{k})^{-1} = \frac{1}{u^2 + x^2 + y^2 + z^2}(u - x\mathbf{i} - y\mathbf{j} - z\mathbf{k}). \tag{8.2}$$

This shows that $\mathbb{H}$ is a skew field.

8.2 Ring Homomorphisms

In this section, we study structure preserving maps between rings, namely ring homomorphisms.

Definition 8.16 A mapping $\varphi : A \to B$ between rings is called a (ring) *homomorphism* if it satisfies the following conditions for any $x, y \in A$:

(1) $\varphi(x + y) = \varphi(x) + \varphi(y)$,
(2) $\varphi(xy) = \varphi(x)\varphi(y)$,
(3) $\varphi(1_A) = 1_B$,

where 1_A and 1_B are unities of A and B, respectively. If there is no risk of confusion, we will write simply 1 in the rest of the text.

Example 8.17 We define $\varphi : \mathbb{C}[X] \to \mathbb{C}$ by

$$\varphi : f(X) \mapsto f(0).$$

It just substitutes X with 0 in each polynomial $f(X)$. For any $f(X), g(X) \in \mathbb{C}[X]$, we have the following:

$$
\begin{aligned}
\varphi(f(X)+g(X)) &= f(0)+g(0) \\
&= \varphi(f(X))+\varphi(g(X)), \\
\varphi(f(X)g(X)) &= f(0)g(0) \\
&= \varphi(f(X))\varphi(g(X)), \\
\varphi(1) &= 1.
\end{aligned}
$$

Thus, φ is a ring homomorphism.

Example 8.18 Let n be a natural number. The mapping $\varphi : \mathbb{Z} \to \mathbb{Z}_n$ given by

$$\varphi : x \mapsto x \mod n$$

is a ring homomorphism.

Example 8.19 The ring $\mathbb{C}[X]$ is an infinite-dimensional complex vector space with a basis $\{1, X, X^2, X^3, \cdots\}$. We consider the linear map $\varphi : \mathbb{C}[X] \to \mathbb{C}[X]$ given by

$$\varphi : \begin{array}{lll} X^n & \mapsto nX^{n-1} & (n > 1) \\ c & \mapsto 0 & c \in \mathbb{C} \end{array},$$

which is the familiar differentiation mapping. We can easily see that φ is not a ring homomorphism. Think of the Leibnitz rule or $\varphi(1) = 0$.

Example 8.20 We revisit Example 8.10. We set

$$A = \{(x, y) \mid x, y \in \mathbb{Z}\}, \quad \text{and} \quad B = \{(x, 0) \mid x \in \mathbb{Z}\}.$$

We define the inclusion map $\varphi : B \to A$ by

$$\varphi : (a, 0) \mapsto (a, 0) \in A.$$

The unity of B is $(1, 0)$ while its image under φ is again $(1, 0)$ which is not the unity of A. So, φ is not qualified to be a ring homomorphism. But, do not jump to a conclusion that this map is useless. It plays an important role later.

The following proposition is derived from Definition 8.16.

Proposition 8.21 *Let $\varphi : A \to B$ be a ring homomorphism. Then, we have the following:*

(1) $\varphi(0_A) = 0_B$.
(2) $\varphi(-a) = -\varphi(a)$.
(3) If $x \in A$ is a unit, then $\varphi(x)$ is also a unit and $\varphi(x)^{-1} = \varphi(x^{-1})$.

Proof Items (1) and (2) means that φ is a group homomorphism between addition groups. They can be easily verified by taking $x = 0$ or $x = -y = a$ in

Definition 8.16. As to (3), from the definition of units, we have the following:

$$x \text{ is a unit } \Leftrightarrow x^{-1} \text{ exists },$$

$$\varphi(x) \text{ is a unit } \Leftrightarrow \varphi(x)^{-1} \text{ exists }.$$

The assumption that x is a unit guarantees the existence of x^{-1}. Then, we have

$$1_B = \varphi(1_A) = \varphi(xx^{-1}) = \varphi(x)\varphi(x^{-1}),$$

which shows that $\varphi(x)^{-1} = \varphi(x^{-1})$. Thus, $\varphi(x)$ is a unit. □

Exercises

[1] Verify that $\mathbb{Z}[\sqrt{-1}] = \{m + n\sqrt{-1} \mid m, n \in \mathbb{Z}\}$ is a ring.
[2] Verify that $\mathbb{Q}[\sqrt{2}] = \{p + q\sqrt{2} \mid p, q \in \mathbb{Q}\}$ is a field.
[3] Show that A below is a ring. Show also that A is a subring of $M_3(\mathbb{C})$.

$$A = \left\{ \begin{pmatrix} a_{11} & a_{12} & 0 \\ a_{21} & a_{22} & 0 \\ 0 & 0 & a_{33} \end{pmatrix} \middle| \; a_{ij} \in \mathbb{C} \right\}$$

[4] Show that B below is a ring. Show also that B is not a subring of $M_3(\mathbb{C})$.

$$B = \left\{ \begin{pmatrix} a_{11} & a_{12} & 0 \\ a_{21} & a_{22} & 0 \\ 0 & 0 & 0 \end{pmatrix} \middle| \; a_{ij} \in \mathbb{C} \right\}$$

[5] We study the ring $\mathbb{Z}_7$.

(1) Write out the tables of addition and multiplication.
(2) Show that $\mathbb{Z}_7^\times = \mathbb{Z}_7 - \{0\}$ is a cyclic group.

[6] We study the ring $\mathbb{Z}_{12}$.

(1) Write out the tables of addition and multiplication.
(2) Pick out all the units and make a multiplication table consisting only these units.
(3) Show that $\mathbb{Z}_{12}^\times \simeq K_4$.

[7] Derive an inverse formula Eq. (8.2). If the reader is unfamiliar with the notion of a cofactor matrix, skip all the guiding hints below and use a brute-force solving method for linear equations. Let

$$A = (a_{ij}) = \begin{pmatrix} u & -x & -y & -z \\ x & u & -z & y \\ y & z & u & -x \\ z & -y & x & u \end{pmatrix}.$$

(Review of linear algebra)
Deleting the i-th row and j-th column from the matrix A, we get a 3×3 matrix A_{ij}. The (i, j)-cofactor is given by $\tilde{a}_{ij} = (-1)^{i+j} \det A_{ij}$.

(1) Calculate the cofactors $\tilde{a}_{11}$, $\tilde{a}_{12}$, $\tilde{a}_{13}$, and $\tilde{a}_{14}$.
(2) Calculate $\det A$.
(Review of linear algebra)
Use the formula

$$\det A = a_{11}\tilde{a}_{11} + a_{12}\tilde{a}_{12} + a_{13}\tilde{a}_{13} + a_{14}\tilde{a}_{14}.$$

(3) If $(u, x, y, z) \neq (0, 0, 0, 0)$, then A^{-1} exists. Calculate its first column.
(Review of linear algebra)
The inverse matrix is $A^{-1} = \frac{1}{\det A}\tilde{A}$, where $\tilde{A}$ is the cofactor matrix of A whose (i, j) entry is the (j, i) cofactor.

$$A^{-1} = \begin{pmatrix} b_{11} & \cdots\cdots\cdots \\ b_{21} & \cdots\cdots\cdots \\ b_{31} & \cdots\cdots\cdots \\ b_{41} & \cdots\cdots\cdots \end{pmatrix}.$$

If we look back Eq. (8.1), we see that the first column of A^{-1} is the solution. Therefore,

$$(u + x\mathbf{i} + y\mathbf{j} + z\mathbf{k})^{-1} = b_{11} + b_{21}\mathbf{i} + b_{31}\mathbf{j} + b_{41}\mathbf{k}$$

is the inversion formula.

Chapter 9
Modules Over Rings

We are already familiar with real and complex vector spaces. The domain of scalars, for example $\mathbb{R}$ or $\mathbb{C}$, brings about various differences between these vector spaces. In general, given a field K, we can consider vector spaces over K, where scalars are taken from K instead of $\mathbb{R}$ or $\mathbb{C}$. Further, we will extend the notion of scalars from fields to rings: from vector spaces to modules over rings.

9.1 Vector Spaces to Modules Over Rings

We start with the definition of vector spaces.

Definition 9.1 Let K be a field. An abelian group V is called a *K-vector space* or *K-linear space* if the following conditions are satisfied:

(1) For any $k \in K$, there is a map $k\bullet : V \to V$ called the *scalar multiplication* by k. We write

$$k\bullet : \vec{v} \mapsto k \bullet \vec{v},$$

where $\bullet$ is often omitted, like $k\vec{v}$, if there is no risk of misunderstanding.

(2) For any $k_1, k_2 \in K$ and $\vec{v} \in V$, we have

$$(k_1 + k_2)\vec{v} = k_1\vec{v} + k_2\vec{v},$$
$$(k_1k_2)\vec{v} = k_1(k_2\vec{v}).$$

(3) For any $k \in K$ and $\vec{v_1}, \vec{v_2} \in V$, we have

$$k(\vec{v_1} + \vec{v_2}) = k\vec{v_1} + k\vec{v_2}.$$

S. Yukita, *First Lectures in Algebra*, Springer Asia Pacific Mathematics Series 10,
https://doi.org/10.1007/978-981-95-5457-7_9

(4) The scalar multiplication of $0, 1 \in K$ with $\vec{v} \in V$ is given as

$$1\vec{v} = \vec{v}, \quad \text{and} \quad 0\vec{v} = \vec{0}.$$

The neutral element of V is denoted by $\vec{0}$ or simply 0 if there is no risk of confusion.

We need homomorphisms between vector spaces.

Definition 9.2 Let K be a field and V, W be K-vector spaces. A map $\varphi : V \to W$ is called a *K-linear map* if it satisfies the following conditions:

(1) $\varphi : V \to W$ is a homomorphism between abelian groups V and W. Namely, for any $\vec{v_1}, \vec{v_2} \in V$ we have

$$\varphi(\vec{v_1} + \vec{v_2}) = \varphi(\vec{v_1}) + \varphi(\vec{v_2}).$$

(2) φ and the scalar multiplication commute. Namely, for any $k \in K$ and $\vec{v} \in V$ we have

$$\varphi(k\vec{v}) = k\varphi(\vec{v}).$$

Remark 9.3 The two conditions in Definition 9.2 can be jointly stated as follows.

($*$) For any $\vec{v_1}, \vec{v_2} \in V$ and $k_1, k_2 \in K$ we have

$$\varphi(k_1\vec{v_1} + k_2\vec{v_2}) = k_1\varphi(\vec{v_1}) + k_2\varphi(\vec{v_2}).$$

Example 9.4 The one-dimensional complex vector space $\mathbb{C}$ can be seen as a two-dimensional real vector space with a basis $\{1, \sqrt{-1}\}$. Any element of $\mathbb{C}$ is uniquely expressed as $x + y\sqrt{-1}$ for some $x, y \in \mathbb{R}$. We define $\varphi : \mathbb{C} \to \mathbb{C}$ by

$$\varphi : x + y\sqrt{-1} \mapsto x + 2y\sqrt{-1},$$

whichi is clearly an $\mathbb{R}$-linear map. The matrix representation with respect to the basis $\{1, \sqrt{-1}\}$ is

$$\begin{pmatrix} 1 & 0 \\ 0 & 2 \end{pmatrix}.$$

In order for φ to be a $\mathbb{C}$-linear map, it must commute with any scalar multiplication by $r + s\sqrt{-1}$ ($r, s \in \mathbb{R}$):

$$\varphi((r + s\sqrt{-1})(x + y\sqrt{-1})) = (r + s\sqrt{-1})\varphi(x + y\sqrt{-1}), \tag{9.1}$$

which does not hold in general. Anyway, $\mathbb{C}$ is a one-dimensional complex vector space, which forces all the $\mathbb{C}$-linear maps to be simple scalar multiplication by complex numbers.

Quiz 9.1 *Show that the $\mathbb{R}$-linear map φ in Example 9.4 is not a $\mathbb{C}$-linear map. It is sufficient to give any vector and scalar that refutes Eq.* (9.1).

Example 9.5 We cannot consider $\mathbb{Z}$ as an $\mathbb{R}$-linear space by any means. Aside from scalar multiplication, we can consider an abelian group $\mathbb{Z}^n$ consisting of n-dimensional column vectors with integer components and with usual componentwise addition. If we restrict scalar multiplication to integers, the abelian group $\mathbb{Z}^n$ is kind of a "vector space" which we will call a $\mathbb{Z}$-linear space or $\mathbb{Z}$-module. Let $\varphi : \mathbb{Z}^2 \to \mathbb{Z}^2$ be defined by

$$\varphi : \begin{pmatrix} a \\ b \end{pmatrix} \mapsto \begin{pmatrix} 1 & 2 \\ 0 & 1 \end{pmatrix} \begin{pmatrix} a \\ b \end{pmatrix} = \begin{pmatrix} a + 2b \\ b \end{pmatrix}.$$

We are tempted to call this map a $\mathbb{Z}$-linear map or homomorphism of $\mathbb{Z}$-module. We will really do so shortly.

Now, we embark on a journey to the world of modules over rings. We extend the notion of scalars from fields to rings.

Definition 9.6 Let A be a ring. An abelian group M is called a *left A-module* if it is equipped with a map

$$\begin{aligned} - \bullet - : & A \times M \to M \\ & (a, x) \mapsto a \bullet x \end{aligned}$$

that satisfies the following conditions:

(1) For any $a_1, a_2 \in A$ and $x \in M$, we have

$$(a_1 + a_2) \bullet x = a_1 \bullet x + a_2 \bullet x.$$

(2) For any $a \in A$ and $x_1, x_2 \in M$, we have

$$a \bullet (x_1 + x_2) = a \bullet x_1 + a \bullet x_2.$$

(3) For any $a_1, a_2 \in A$ and $x \in M$, we have

$$(a_1 a_2) \bullet x = a_1 \bullet (a_2 \bullet x).$$

(4) For $0, 1 \in A$ and any $x \in M$, we have

$$1 \bullet x = x, \quad \text{and} \quad 0 \bullet x = 0_M,$$

where 0_M denotes the neutral element of M to distinguish from $0 \in A$. We often omit the subscript where there is no risk of confusion.

For any $a \in A$, the map $a\bullet : M \to M$ called the *left scalar multiplication* by a. We write

$$a\bullet : x \mapsto a \bullet x,$$

where $\bullet$ is often omitted, like ax, if there is no risk of misunderstanding.

The left A-module M may be expressed as $(M, \bullet)$ to make clear the scalar multiplication involved. Then, M may be called the *underlying additive group* or *underlying abelian group*.

If a ring A is commutative such as $\mathbb{Z}$, $\mathbb{Q}$, $\mathbb{R}$, and $\mathbb{C}$, left A-modules are simply called A-modules dropping the word "left." Real and complex vector spaces may be called $\mathbb{R}$-modules and $\mathbb{C}$-modules, respectively.

All abelian groups are regarded as $\mathbb{Z}$-modules. Let M be an abelian group. Let $n \in \mathbb{Z}$ and $x \in M$. The scalar multiplication by n is defined as follows:

$$n \bullet x = \begin{cases} \underbrace{x + x + \cdots + x}_{n \text{ times}} & (\text{if } n > 0) \\ 0_M & (\text{if } n = 0) \\ \underbrace{-x - x - \cdots - x}_{|n| \text{ times}} & (\text{if } n < 0) \end{cases}$$

This satisfies all the conditions in Definition 9.6.

Quiz 9.2 *Prove this.*

There is no other choice of $\mathbb{Z}$ scalar multiplication.

Proposition 9.7 *Given an abelian group M, the $\mathbb{Z}$-module structure is unique. In other words, if there is another scalar multiplication, say $n :$, we have $n : x = n \bullet x$ for all $n \in \mathbb{Z}$ and $x \in M$.*

Proof Condition (4) in Definition 9.6 gives

$$1 : x = x = 1 \bullet x.$$

For $n > 0$, if we use condition (1) repeatedly, then we have

$$n : x = \underbrace{(1 + 1 + \cdots + 1)}_{n \text{ times}} : x = \underbrace{x + x + \cdots + x}_{n \text{ times}} = \underbrace{(1 + 1 + \cdots + 1)}_{n \text{ times}} \bullet x = n \bullet x.$$

Likewise, we can verify the case of $n \leq 0$. □

If a ring A is noncommutative such as $M_n(\mathbb{Z})$, $M_n(\mathbb{C})$, and $\mathbb{H}$, the adjectives "left" and "right" make a difference. We will see right modules later.

Typical examples of left modules are given below.

Example 9.8 We consider the ring of $n \times n$ complex matrices $A = M_n(\mathbb{C})$. Let $M = \mathbb{C}^n$ be an n-dimensional column vector space with components in $\mathbb{C}$. The left scalar multiplication of a module element $\begin{pmatrix} x_1 \\ x_2 \\ \vdots \\ x_n \end{pmatrix} \in M$ and a scalar $\begin{pmatrix} a_{11} & a_{12} & \cdots & a_{1n} \\ a_{21} & a_{22} & \cdots & a_{2n} \\ \vdots & \vdots & \ddots & \vdots \\ a_{n1} & a_{n2} & \cdots & a_{nn} \end{pmatrix} \in A$ is defined by the usual matrix multiplication as follows:

$$\begin{pmatrix} a_{11} & a_{12} & \cdots & a_{1n} \\ a_{21} & a_{22} & \cdots & a_{2n} \\ \vdots & \vdots & \ddots & \vdots \\ a_{n1} & a_{n2} & \cdots & a_{nn} \end{pmatrix} \begin{pmatrix} x_1 \\ x_2 \\ \vdots \\ x_n \end{pmatrix}.$$

With this scalar multiplication, the abelian group M is a left A-module.

Another example of a left module is given below.

Example 9.9 We consider the ring of $n \times n$ complex matrices $A = M_n(\mathbb{C})$. Let $M = M_{n,2}(\mathbb{C})$ be an abelian group of $n \times 2$ matrices under matrix addition with components in $\mathbb{C}$. The left scalar multiplication of a module element $\begin{pmatrix} x_{11} & x_{12} \\ x_{21} & x_{22} \\ \vdots & \vdots \\ x_{n1} & x_{n2} \end{pmatrix} \in M$ and a scalar $\begin{pmatrix} a_{11} & a_{12} & \cdots & a_{1n} \\ a_{21} & a_{22} & \cdots & a_{2n} \\ \vdots & \vdots & \ddots & \vdots \\ a_{n1} & a_{n2} & \cdots & a_{nn} \end{pmatrix} \in A$ is defined by the usual matrix multiplication as follows:

$$\begin{pmatrix} a_{11} & a_{12} & \cdots & a_{1n} \\ a_{21} & a_{22} & \cdots & a_{2n} \\ \vdots & \vdots & \ddots & \vdots \\ a_{n1} & a_{n2} & \cdots & a_{nn} \end{pmatrix} \begin{pmatrix} x_{11} & x_{12} \\ x_{21} & x_{22} \\ \vdots & \vdots \\ x_{n1} & x_{n2} \end{pmatrix}.$$

With this scalar multiplication, the abelian group M is a left A-module.

Recall what we have discussed in Chap. 1 concerning the matrix algebra. The matrix multiplication from the left is performed columnwise in the right applicand.

The matrix multiplication between an $n \times n$ matrix and $n \times 2$ matrix is understood as a combination of the two multiplications

$$\begin{pmatrix} a_{11} & a_{12} & \cdots & a_{1n} \\ a_{21} & a_{22} & \cdots & a_{2n} \\ \vdots & \vdots & \ddots & \vdots \\ a_{n1} & a_{n2} & \cdots & a_{nn} \end{pmatrix} \begin{pmatrix} x_{11} \\ x_{21} \\ \vdots \\ x_{n1} \end{pmatrix} \text{ and } \begin{pmatrix} a_{11} & a_{12} & \cdots & a_{1n} \\ a_{21} & a_{22} & \cdots & a_{2n} \\ \vdots & \vdots & \ddots & \vdots \\ a_{n1} & a_{n2} & \cdots & a_{nn} \end{pmatrix} \begin{pmatrix} x_{12} \\ x_{22} \\ \vdots \\ x_{n2} \end{pmatrix}.$$

In this way, we can find more examples of left modules.

The next remark is for advanced readers.

Remark 9.10 Given two endomorphisms φ and ψ of an abelian group M, we define addition $\varphi + \psi$ and multiplication $\varphi\psi$ by

$$(\varphi + \psi)(x) = \varphi(x) + \psi(x) \quad \text{and}$$
$$(\varphi\psi)(x) = \varphi(\psi(x)).$$

With this addition and multiplication, $\mathrm{End}(M)$ is qualified to be a ring, where the zero element is the endomorphism that sends the whole M to 0_M, and the unity is the identity endomorphism. Definition 9.6 says that an A-module is an abelian group M with a ring homomorphism $A \to \mathrm{End}(M)$.

We are naturally led to the following definition for homomorphisms between modules.

Definition 9.11 Let M and N be left A-modules. A homomorphism $\varphi : M \to N$ between underlying additive groups is called a *left A-module homomorphism* or *left A-linear map* if the following conditions are satisfied:

(*) The scalar multiplication and φ commute. Namely, for any $a \in A$ and $x \in M$ we have

$$\varphi(ax) = a\varphi(x).$$

In group theory, the notions of subgroups and quotients/factors play important roles. We introduce corresponding notions in module theory.

Definition 9.12 Let M be a left A-module. A subgroup N of the underlying additive group is called a *left A-submodule* of M if it satisfies the following condition:

(*) For any $a \in A$ and $x \in N$, we have $ax \in N$. In other words, N is closed under the left scalar multiplication.

The trivial subgroups $\{0\}$ and M are left A-submodules. These are called *trivial submodules*.

So much for left modules. Next, we consider right modules.

Definition 9.13 Let A be a ring. An abelian group M is called a *right A-module* if it is equipped with a map

$$
\begin{aligned}
- \bullet - : & M \times A \to M \\
& (x, a) \mapsto x \bullet a
\end{aligned}
$$

that satisfies the following conditions:

(1) For any $a_1, a_2 \in A$ and $x \in M$, we have

$$x \bullet (a_1 + a_2) = x \bullet a_1 + x \bullet a_2.$$

(2) For any $a \in A$ and $x_1, x_2 \in M$, we have

$$(x_1 + x_2) \bullet a = x_1 \bullet a + x_2 \bullet a.$$

(3) For any $a_1, a_2 \in A$ and $x \in M$, we have

$$x \bullet (a_1 a_2) = (x \bullet a_1) \bullet a_2.$$

(4) For $0, 1 \in A$ and any $x \in M$, we have

$$x \bullet 1 = x, \quad \text{and} \quad x \bullet 0 = 0_M,$$

where 0_M denotes the neutral element of M to distinguish from $0 \in A$. We often omit the subscript where there is no risk of confusion.

For any $a \in A$, the map $\bullet a : M \to M$ called the *right scalar multiplication* by a. We write

$$\bullet a : x \mapsto x \bullet a,$$

where $\bullet$ is often omitted, like xa, if there is no risk of misunderstanding.

Typical examples of right modules are given below.

Example 9.14 We consider the ring of $n \times n$ complex matrices $A = M_n(\mathbb{C})$. Let $M = \mathbb{C}^n$ be an n-dimensional row vector space with components in $\mathbb{C}$. Let $\begin{pmatrix} x_1 & x_2 & \cdots & x_n \end{pmatrix} \in M$. The right scalar multiplication by $\begin{pmatrix} a_{11} & a_{12} & \cdots & a_{1n} \\ a_{21} & a_{22} & \cdots & a_{2n} \\ \vdots & \vdots & \ddots & \vdots \\ a_{n1} & a_{n2} & \cdots & a_{nn} \end{pmatrix}$ is defined by the usual matrix multiplication as follows:

$$(x_1\ x_2 \cdots x_n)\begin{pmatrix} a_{11} & a_{12} & \cdots & a_{1n} \\ a_{21} & a_{22} & \cdots & a_{2n} \\ \vdots & \vdots & \ddots & \vdots \\ a_{n1} & a_{n2} & \cdots & a_{nn} \end{pmatrix}.$$

With this scalar multiplication, the abelian group M is a right A-module.

Another example of a right module is given below.

Example 9.15 We consider the ring of $n \times n$ complex matrices $A = M_n(\mathbb{C})$. Let $M = M_{2,n}(\mathbb{C})$ be an abelian group of $2 \times n$ matrices under matrix addition with components in $\mathbb{C}$. The right scalar multiplication of a module element $\begin{pmatrix} x_{11} & x_{12} & \cdots & x_{1n} \\ x_{21} & x_{22} & \cdots & x_{2n} \end{pmatrix} \in M$ and a scalar $\begin{pmatrix} a_{11} & a_{12} & \cdots & a_{1n} \\ a_{21} & a_{22} & \cdots & a_{2n} \\ \vdots & \vdots & \ddots & \vdots \\ a_{n1} & a_{n2} & \cdots & a_{nn} \end{pmatrix} \in A$ is defined by the usual matrix multiplication as follows:

$$\begin{pmatrix} x_{11} & x_{12} & \cdots & x_{1n} \\ x_{21} & x_{22} & \cdots & x_{2n} \end{pmatrix}\begin{pmatrix} a_{11} & a_{12} & \cdots & a_{1n} \\ a_{21} & a_{22} & \cdots & a_{2n} \\ \vdots & \vdots & \ddots & \vdots \\ a_{n1} & a_{n2} & \cdots & a_{nn} \end{pmatrix}.$$

With this scalar multiplication, the abelian group M is a right A-module.

Recall what we have discussed in Chap. 1 concerning the matrix algebra. Recall also Example 9.9 and verify the following.

The matrix multiplication from the right is performed row by row in the left applicand. The matrix multiplication between a $2 \times n$ matrix and $n \times n$ matrix is understood as a combination of the two multiplications between a row vector and a square matrix. In this way, we can find more examples of right modules.

Finally, we introduce the concept of bimodules.

Definition 9.16 Let A and B be rings. An abelian group M is called a *A-B-bimodule* if it satisfies the following two conditions:

(1) M is a left A-module with a left scalar multiplication $\bullet$ and a right B-module with a right scalar multiplication $*$.
(2) For all $a \in A$, $b \in B$, and $m \in M$,

$$(a \bullet m) * b = a \bullet (m * b).$$

We often omit $\bullet$ and $*$, like amb, if there is no risk of misunderstanding. When $A = B$, we simply say that M is an A-bimodule.

Example 9.17 We consider a 6-dimensional complex vector space

$$M = \left\{ \begin{pmatrix} m_{11} & m_{12} & m_{13} \\ m_{21} & m_{22} & m_{23} \end{pmatrix} \middle| m_{ij} \in \mathbb{C} \right\}.$$

The abelian group M is a $M_2(\mathbb{C})$-$M_3(\mathbb{C})$-bimodule. Generalization is obvious.

Let us return to left modules and consider their left submodules.

Example 9.18 We consider the ring $A = M_2(\mathbb{R})$ and the left A-module $M = \mathbb{R}^2$, where $\mathbb{R}^2$ is the two-dimensional column vector space. Let N be a subgroup of M given by

$$N = \left\{ \begin{pmatrix} z \\ 0 \end{pmatrix} \middle| z \in \mathbb{R} \right\}.$$

The subgroup N is a $\mathbb{R}$-submodule of M. However, N is not a left A-submodule of M. Consider the following scalar multiplication for $z \neq 0$.

$$\begin{pmatrix} 0 & 1 \\ 1 & 0 \end{pmatrix} \begin{pmatrix} z \\ 0 \end{pmatrix} = \begin{pmatrix} 0 \\ z \end{pmatrix} \notin N.$$

This shows that N is not closed under the left scalar multiplication by A. Further, we can show that there is no left A-submodule of M except for trivial submodules $\{0\}$ and M. Suppose N is a nontrivial left A-submodule of M. Then, there must be nonzero elements $\begin{pmatrix} a \\ b \end{pmatrix}$ and $\begin{pmatrix} c \\ d \end{pmatrix}$ such that

$$\begin{pmatrix} a \\ b \end{pmatrix} \in N \quad \text{and} \quad \begin{pmatrix} c \\ d \end{pmatrix} \notin N.$$

There is a third vector $\begin{pmatrix} p \\ q \end{pmatrix}$ that is linearly independent with $\begin{pmatrix} a \\ b \end{pmatrix}$ and $\begin{pmatrix} c \\ d \end{pmatrix}$ at the same time. Take, for example $\begin{pmatrix} p \\ q \end{pmatrix} = \begin{pmatrix} a+c \\ b+d \end{pmatrix}$. We seek a 2×2 matrix X such that

$$X \begin{pmatrix} a \\ b \end{pmatrix} = \begin{pmatrix} c \\ d \end{pmatrix} \quad \text{and} \quad X \begin{pmatrix} p \\ q \end{pmatrix} = \begin{pmatrix} p \\ q \end{pmatrix}.$$

If we combine these equations, we have a single matrix equation

$$X\begin{pmatrix} a & p \\ b & q \end{pmatrix} = \begin{pmatrix} c & p \\ d & q \end{pmatrix}.$$

Thus, we have

$$X = \begin{pmatrix} c & p \\ d & q \end{pmatrix}\begin{pmatrix} a & p \\ b & q \end{pmatrix}^{-1}.$$

Since $X\begin{pmatrix} a \\ b \end{pmatrix} = \begin{pmatrix} c \\ d \end{pmatrix}$ for $X \in A$, we conclude that N is not closed under scalar multiplication, which contradicts the assumption.

Remark 9.19 In Example 9.18 we used the techniques for matrix construction introduced in Sect. 1.5.

Example 9.20 Let A be the ring of diagonal 2×2-matrices with components in $\mathbb{C}$:

$$A = \left\{ \begin{pmatrix} a_1 & 0 \\ 0 & a_2 \end{pmatrix} \middle| a_1, a_2 \in \mathbb{C} \right\}.$$

The ring A is commutative. The column vector space $M = \mathbb{C}^2$ can be seen as a left A-module. Consider its subgroups as $\mathbb{C}$ vector spaces below. and its subspaces

$$N_1 = \left\{ \begin{pmatrix} z \\ 0 \end{pmatrix} \middle| z \in \mathbb{C} \right\},$$

$$N_2 = \left\{ \begin{pmatrix} 0 \\ z \end{pmatrix} \middle| z \in \mathbb{C} \right\},$$

$$N_3 = \left\{ \begin{pmatrix} z \\ z \end{pmatrix} \middle| z \in \mathbb{C} \right\}.$$

N_1 and N_2 are A-submodules of M, whereas N_3 is not. In fact, take $\begin{pmatrix} 1 & 0 \\ 0 & -1 \end{pmatrix} \in A$. We have

$$\begin{pmatrix} 1 & 0 \\ 0 & -1 \end{pmatrix}\begin{pmatrix} z \\ z \end{pmatrix} = \begin{pmatrix} z \\ -z \end{pmatrix} \notin N_3$$

for $z \neq 0$.

Quiz 9.3 *Verify that N_1 and N_2 are A-submodule of M.*

Example 9.21 Let A be the non-commutative ring given below:

$$A = \left\{ \begin{pmatrix} a_{11} & a_{12} & 0 \\ a_{21} & a_{22} & 0 \\ 0 & 0 & a_{33} \end{pmatrix} \middle| a_{ij} \in \mathbb{C} \right\}.$$

Let $M = \mathbb{C}^3$ be the column vector space. We consider a left A-module $M = \mathbb{C}^3$. All A-submodules of M are

$$N_1 = \left\{ \begin{pmatrix} x \\ y \\ 0 \end{pmatrix} \middle| x, y \in \mathbb{C} \right\} \quad \text{and}$$

$$N_2 = \left\{ \begin{pmatrix} 0 \\ 0 \\ z \end{pmatrix} \middle| z \in \mathbb{C} \right\},$$

except for trivial submodules.

Quiz 9.4 *Show that N_1 and N_2 are left A-modules of M.*

Quiz 9.5 *Show that any left A-submodule of M containing $\begin{pmatrix} a \\ b \\ 0 \end{pmatrix} \neq \begin{pmatrix} 0 \\ 0 \\ 0 \end{pmatrix}$ contains the whole N_1. In other words, the A-submodule N_1 is generated by $\begin{pmatrix} a \\ b \\ 0 \end{pmatrix} \neq \begin{pmatrix} 0 \\ 0 \\ 0 \end{pmatrix}$. Show also that any left A-submodule of M containing $\begin{pmatrix} 0 \\ 0 \\ z \end{pmatrix} \neq \begin{pmatrix} 0 \\ 0 \\ 0 \end{pmatrix}$ contains the whole N_2.*

Quizzes 9.4 and 9.5 reveal that N_1 and N_2 are the only left A-submodules of M except for trivial ones.

Remark 9.22 Given any ring A, we can regard A as a left or right A-module or A-bimodule. Based on this, the notion of ideals will be introduced in the next chapter.

Let us see some examples of module homomorphisms.

Example 9.23 Let A be a commutative ring

$$A = \left\{ \begin{pmatrix} a_1 & 0 \\ 0 & a_2 \end{pmatrix} \middle| a_1, a_2 \in \mathbb{C} \right\}.$$

Let $M = \mathbb{C}^2$ be a left A-module consisting of 2-dimensional column vectors. We define a complex linear map $\varphi : M \to M$ as follows:

$$\varphi : \begin{pmatrix} x \\ y \end{pmatrix} \mapsto \begin{pmatrix} 0 & 1 \\ 1 & 0 \end{pmatrix} \begin{pmatrix} x \\ y \end{pmatrix} = \begin{pmatrix} y \\ x \end{pmatrix}.$$

Is φ an A-module homomorphism? Let us see if φ commutes with A-scalar multiplication:

$$\begin{pmatrix} 0 & 1 \\ 1 & 0 \end{pmatrix}\begin{pmatrix} a_1 & 0 \\ 0 & a_2 \end{pmatrix} = \begin{pmatrix} a_1 & 0 \\ 0 & a_2 \end{pmatrix}\begin{pmatrix} 0 & 1 \\ 1 & 0 \end{pmatrix}.$$

This does not hold in general.

We define a complex linear map $\psi : M \to M$ as follows:

$$\psi : \begin{pmatrix} x \\ y \end{pmatrix} \mapsto \begin{pmatrix} 2 & 0 \\ 0 & 3 \end{pmatrix}\begin{pmatrix} x \\ y \end{pmatrix} = \begin{pmatrix} 2x \\ 3y \end{pmatrix}.$$

Is ψ an A-module homomorphism? Let us see if ψ commutes with A-scalar multiplication:

$$\begin{pmatrix} 2 & 0 \\ 0 & 3 \end{pmatrix}\begin{pmatrix} a_1 & 0 \\ 0 & a_2 \end{pmatrix} = \begin{pmatrix} a_1 & 0 \\ 0 & a_2 \end{pmatrix}\begin{pmatrix} 2 & 0 \\ 0 & 3 \end{pmatrix},$$

which is shown to be true by simple calculation.

Quiz 9.6 *We have seen two $\mathbb{C}$-linear maps φ and ψ in Example 9.23, Refute the claim that φ is a left A-module homomorphism. Prove that ψ is a left A-module homomorphism.*

Remark 9.24 (Reinforced Concrete) A module over a ring can be considered as an object made of reinforced concrete, where reinforcement is done by a ring (iron rod), the concrete is an abelian group. Given a left A-module M, consider the set

$$Ax = \{ax \mid a \in A\}$$

for every $x \in M$. We may call it the *scalar rebar* (iron) that passes through x in the concrete (abelian group). Note that this analogy and terminology is not used in standard textbooks. The reader is kindly requested to distribute this idea to math enthusiasts.

Let $\varphi : M \to N$ be a left A-module homomorphism. The map φ sends a scalar rebar in M to a scalar rebar in N. See Fig. 9.1.

Fig. 9.1 Rebar to rebar

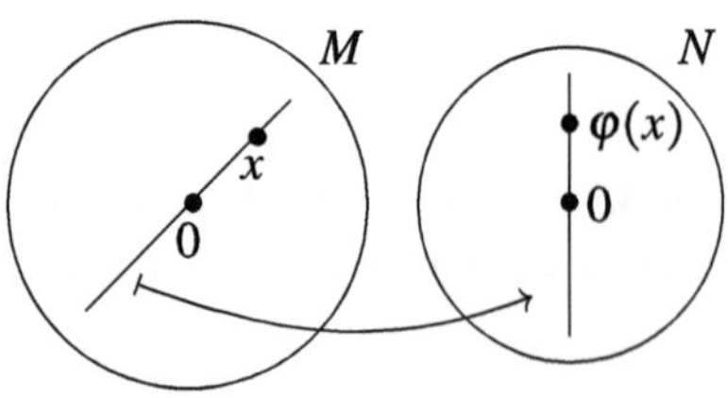

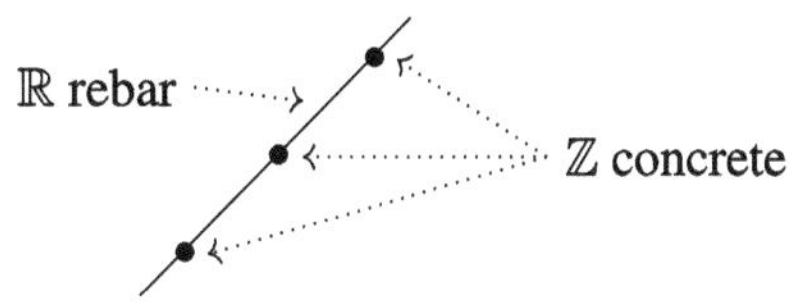

Fig. 9.2 A forcible $\mathbb{R}$ rebar protrudes $\mathbb{Z}$ concrete

A $\mathbb{Z}$-module is a concrete object without reinforcement. An $\mathbb{R}$-module is a concrete object with a light reinforcement. A $\mathbb{C}$-module is a concrete object with a heavy reinforcement. A left $M_2(\mathbb{C})$-module a concrete object with a grid reinforcement. An abelian group $\mathbb{Z}$ cannot be reinforced by the ring $\mathbb{R}$ since the $\mathbb{R}$-rebars protrude from the concrete. See Fig. 9.2.

9.2 Module Homomorphisms and Factor Modules

We can discuss module homomorphisms and factors/quotients in parallel with those of groups.

Theorem 9.25 *Let $\varphi : M \to N$ be a left A-module homomorphism. The map φ has, as a homomorphism between underlying additive groups, the kernel and image. We have the following:*

(1) *The additive subgroup* $\ker \varphi$ *is a left A-submodule of M.*
(2) *The additive subgroup* $\operatorname{Im} \varphi$ *is a left A-submodule of N.*

Proof We show that both $\ker \varphi$ and $\operatorname{Im} \varphi$ are closed under scalar multiplication.

(1) For any scalar $a \in A$ and element $x \in \ker \varphi$, the property of the left A-homomorphism gives the following:

$$\varphi(ax) = a\varphi(x) = a0_N = 0_N.$$

This shows that $ax \in \ker \varphi$.
(2) For any scalar $a \in A$ and element $y \in \operatorname{Im} \varphi$, we have:

- $y = \varphi(x)$ for some $x \in M$.
- $\varphi(ax) = a\varphi(x) = ay$.

Therefore, we have $ay \in \operatorname{Im} \varphi$.

□

Theorem 9.25 allows the following definition.

Definition 9.26 Let $\varphi : M \to N$ be a left A-module homomorphism. The *kernel* of φ is the kernel of the underlying group homomorphism equipped with the left A-module structure and denoted also by $\ker \varphi$.

The *image* of φ is the image of the underlying group homomorphism equipped with the left A-module structure and denoted also by $\operatorname{Im}\varphi$.

We reproduce what we have discussed in Sect. 4.2 in the context of modules over rings. Given a left A-module homomorphism $\varphi : M \to N$, we can consider the factor group $M/\ker\varphi$ of the underlying additive groups.

Theorem 9.27 *Let L be a left A-submodule of M. The factor group M/L of the underlying additive groups is a left A-module. Every L-orbit is closed under scalar multiplication. The canonical projection map $\pi : M \to M/L$ is a left A-module homomorphism.*

Proof (The First Half) The subgroup L of M is automatically normal since M is abelian. Therefore, we can talk about the factor group M/L. We express an element of M/L as $x + L$ with a representative $x \in M$. Since L is a left A-submodule, we have $aL \subset L$. Thus, we have

$$a(x + L) \subset ax + L.$$

Let $y \in ax + L$. We have $a^{-1}y \in x + a^{-1}L \subset x + L$, consequently $y \in a(x + L)$. Thus, we have

$$a(x + L) = ax + aL,$$

which means that scalar multiplication sends any L-orbit to another. We can introduce scalar multiplication $\bullet$ to the set of L-orbits:

$$a \bullet (x + L) = ax + L.$$

This operation $\bullet$ is well-defined. Take any other representative x'. We must have $x' - x \in L$ and

$$\begin{aligned} a \bullet (x' + L) &= a \bullet (x + L) + a \bullet (x' - x + L) \\ &= ax + L + aL \\ &= ax + L, \end{aligned}$$

where we used $aL \subset L$ therefore $L + aL = L$.

To show that the abelian group M/L a left A-module with this operator $\bullet$, we have to verify the following conditions:

(1) For any $a_1, a_2 \in A$ and $x + L \in M/L$, we have

$$(a_1 + a_2) \bullet (x + L) = a_1 \bullet (x + L) + a_2 \bullet (x + L).$$

(2) For any $a \in A$ and $x_1 + L, x_2 + L \in M/L$, we have

$$a \bullet ((x_1 + L) + (x_2 + L)) = a \bullet (x_1 + L) + a \bullet (x_2 + L).$$

(3) For any $a_1, a_2 \in A$ and $x + L \in M/L$, we have

$$(a_1 a_2) \bullet (x + L) = a_1 \bullet (a_2 \bullet (x + L)).$$

(4) For $0, 1 \in A$ and any $x + L \in M/L$, we have

$$1 \bullet (x + L) = x + L, \quad 0 \bullet (x + L) = 0 + L.$$

Note that $0 + L = 0_{M/L}$.

We leave the details of (1)–(4) to the quiz below. □

Quiz 9.7 *Verify conditions (1)–(4) in Theorem 9.27. Freely use the rewriting rule* $(x_1 + L) + (x_2 + L) = (x_1 + x_2) + L$.

Proof (The Second Half of 9.27) We prove that the underlying group homomorphism $\pi : M \to M/L$ is a left A-module homomorphism. This amounts to showing that the following diagram commutes:

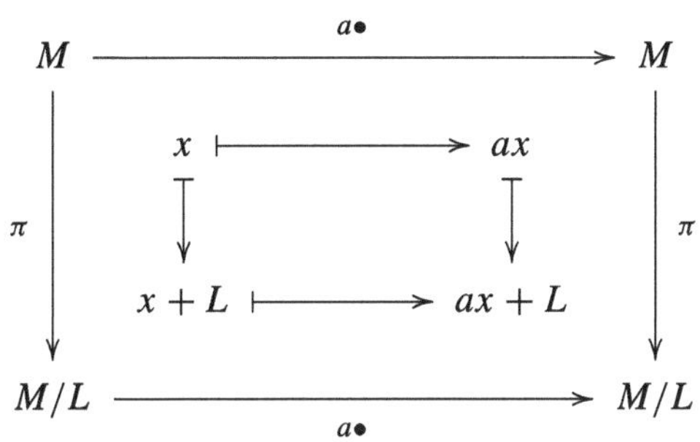

We do it element by element. Let x be any element of M at the upper left corner. Calculation shows the following:

$$a \cdot \pi(x) = a \cdot (x + L) = ax + L,$$
$$\pi(a \cdot x) = \pi(ax) = ax + L.$$

□

Theorem 9.27 allows the following definition.

Definition 9.28 Let A be a ring. Let N be a left A-submodule of a left A-module M. The factor group M/N of the underlying groups can be uniquely seen as a left A-module and is called the *factor module* or *quotient module*. The canonical projection map $\varphi : M \to M/N$ of the underlying abelian groups is a left A-module homomorphism and is also called the *canonical projection map* or *canonical projection homomorphism*.

Theorem 9.29 *Let $\varphi : M \to N$ be a left A-module homomorphism. We have the following.*

(*) *There is a unique left A-module isomorphism $\tilde{\varphi}$ that makes the following diagram commute:*

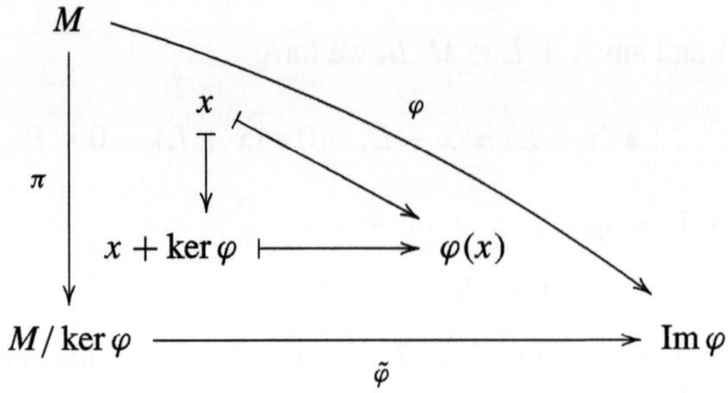

Proof For the underlying groups and homomorphisms, by Theorem 6.3 we have a group isomorphism $\tilde{\varphi}$ defined by

$$\tilde{\varphi}(x + \ker \varphi) = \varphi(x).$$

It remains to prove that $\tilde{\varphi}$ commutes with the scalar multiplication. Let us see:

$$\begin{aligned} & \tilde{\varphi}(a \cdot (x + \ker \varphi)) \\ = \; & \tilde{\varphi}(ax + \ker \varphi) \\ = \; & \varphi(ax) = a\varphi(x) \\ = \; & a\tilde{\varphi}(x + \ker \varphi). \end{aligned}$$

□

Remark 9.30 Let $\varphi : G \to H$ be a homomorphism between abelian groups. By Proposition 9.7, we may regard φ as a $\mathbb{Z}$-module homomorphism.

Example 9.31 Let us define $\varphi : \mathbb{Z} \to \mathbb{Z}$ by $\varphi(n) = 2n$. The map φ is a $\mathbb{Z}$-module homomorphism with the kernel and image as follows:

$$\begin{aligned} \ker \varphi &= \{0\}, \\ \operatorname{Im} \varphi &= 2\mathbb{Z} = \{2n \mid n \in \mathbb{Z}\}. \end{aligned}$$

The $\mathbb{Z}$-module isomorphism $\tilde{\varphi} : \mathbb{Z}/\ker \varphi \to \operatorname{Im} \varphi$ is given by

$$n + \{0\} \mapsto 2n.$$

Example 9.32 Let us define a ring A by

$$A = \left\{ \begin{pmatrix} a_{11} & a_{12} & 0 \\ a_{21} & a_{22} & 0 \\ 0 & 0 & a_{33} \end{pmatrix} \middle| \, a_{ij} \in \mathbb{C} \right\}.$$

The column vector space $\mathbb{C}^3$ is regarded as a left A-module. Elements of A that commute all other elements must be

$$\begin{pmatrix} a & 0 & 0 \\ 0 & a & 0 \\ 0 & 0 & b \end{pmatrix}, \quad a, b \in \mathbb{C}.$$

Take one of such matrices and define $\varphi : \mathbb{C}^3 \to \mathbb{C}^3$ by

$$\varphi : \begin{pmatrix} x \\ y \\ z \end{pmatrix} \mapsto \begin{pmatrix} a & 0 & 0 \\ 0 & a & 0 \\ 0 & 0 & b \end{pmatrix} \begin{pmatrix} x \\ y \\ z \end{pmatrix}.$$

Then, φ is a left A-module homomorphism.

Example 9.33 Let k be any field. Consider the polynomial ring $k[X]$. The field k is regarded as a $k[X]$-module. We define the scalar multiplication $f(X) \bullet z$ between the scalar $f(X)$ and a module element $z \in k$ by $f(0)z$.

We define a $k[X]$-module homomorphism $\varphi : k[X] \to k$ by

$$\varphi : f(X) \mapsto f(0).$$

Then, we have the following:

$$\ker \varphi = X\,k[X] = \{Xf(X) \mid f(X) \in k[X]\},$$
$$\operatorname{Im} \varphi = k.$$

See exercises.

Quiz 9.8 *Verify that $X\,k[X]$ is a $k[X]$-submodule of $k[X]$. Prove that φ is surjective.*

The map φ in Example 9.33 is a ring homomorphism. Revisit Example 8.17.

9.3 Phenomena of Modules Over Rings

The notion of modules over rings is a generalization of the notion of vector spaces. This generalization provides us with new phenomena that are forbidden by vector spaces.

We begin by reviewing vector spaces. Let k be a field and W a k-vector space. If there are k-subspaces U and V such that

$$U \cap V = \{\vec{0}\}, \quad W = U + V \stackrel{\text{def}}{=} \{\vec{u} + \vec{v} \mid \vec{u} \in U, \vec{v} \in V\},$$

we say that V is a *complementary subspace* of U or vice versa. We write

$$W = U \oplus_k V$$

and say W is decomposed into the *direct sum* of U and V. We often omit k and simply write $W = U \oplus V$ when there is no risk of misunderstanding.

It is easy to see that every subspace U of a finite-dimensional k-vector space W has a complementary subspace. Let $\vec{e_1}, \ldots, \vec{e_n}$ be a basis of U. Extending the basis of U, we can find a basis of W such as $\vec{e}_1, \ldots, \vec{e}_n, \vec{e}_{n+1}, \ldots, \vec{e}_m$. Let V be a subspace of W spanned by $\vec{e}_{n+1}, \ldots, \vec{e}_m$. Then, V is a subspace complementary to U.

Remark 9.34 The discussion using a basis can be done with infinite-dimensional cases. See advanced textbooks or articles on the Internet.

Example 9.35 Let $W = \mathbb{R}^2$ be the column vector space. Let U and V be subspaces defined by

$$U = \left\{ \begin{pmatrix} x \\ 0 \end{pmatrix} \middle| x \in \mathbb{R} \right\}$$

and

$$V = \left\{ \begin{pmatrix} 0 \\ y \end{pmatrix} \middle| y \in \mathbb{R} \right\}.$$

V is a complementary subspace of U. Notice that there are many other complementary subspaces of U. For example, the following V_2 is complementary to U:

$$V_2 = \left\{ \begin{pmatrix} y \\ y \end{pmatrix} \middle| y \in \mathbb{R} \right\}.$$

Quiz 9.9 *Verify that* $W = U \oplus V_2$.

Now, we discuss complementary submodules.

Definition 9.36 Let U and V be left A-submodules of W. If they satisfy the conditions

$$U \cap V = \{0_W\} \quad \text{and} \quad W = U + V \stackrel{\text{def}}{=} \{u + v \mid u \in U, v \in V\},$$

we write

$$W = U \oplus V.$$

We say V is a *complementary submodule* or simply *complement* of U, and vice versa. The expression $U \oplus V$ is called a *direct sum* of A-modules, where U and V are called *direct summands*.

Unlike vector spaces, complementary submodules do not always exist.

Example 9.37 Recall that any $\mathbb{Z}$-submodule of $\mathbb{Z}$ can be expressed as $n\mathbb{Z}$ for some natural number n. (See Theorem 1.6 or exercises in Chap. 3.)

Let us consider $\mathbb{Z}$-submodule $2\mathbb{Z}$ of $\mathbb{Z}$. We wonder if its complement exists or not. Suppose

$$\mathbb{Z} = 2\mathbb{Z} + n\mathbb{Z}.$$

The necessary and sufficient condition for n is that n and 2 are coprimes. Let n be any natural number coprime with 2. Then, we have

$$2n \in 2\mathbb{Z} \cap n\mathbb{Z}.$$

So, we cannot attain the goal $2\mathbb{Z} \cap n\mathbb{Z} = \{0\}$. Thus, $2\mathbb{Z}$ has no complements.

The example above shows that modules over rings are different from vector spaces.

The following example shows how complementary submodules come into existence.

Example 9.38 We consider the ring $\mathbb{Z}_6 = \{\bar{0}, \bar{1}, \ldots, \bar{5}\}$ with modular arithmetic modulo 6. We consider $\mathbb{Z}_6$-submodules

$$2\mathbb{Z}_6 = \{\bar{0}, \bar{2}, \bar{4}\} \simeq \mathbb{Z}_3 \quad \text{and}$$

$$3\mathbb{Z}_6 = \{\bar{0}, \bar{3}\} \simeq \mathbb{Z}_2.$$

We omit bars for calculation below.

We prove that $2\mathbb{Z}_6 \oplus 3\mathbb{Z}_6 = \mathbb{Z}_6$. The round-robin addition table is given below.

$2\mathbb{Z}_6$	$3\mathbb{Z}_6$	$2\mathbb{Z}_6 + 3\mathbb{Z}_6$
0	0	0
0	3	3
2	0	2
2	3	5
4	0	4
4	3	1

This shows that

$$\mathbb{Z}_6 = 2\mathbb{Z}_6 + 3\mathbb{Z}_6 \quad \text{and} \quad 2\mathbb{Z}_6 \cap 3\mathbb{Z}_6 = \{0\}.$$

We verify that $2\mathbb{Z}_6$ and $3\mathbb{Z}_6$ are $\mathbb{Z}_6$-submodules. The tables of scalar multiplications $2\times$ and $3\times$ are given below.

$2\times$			$3\times$		
$2\mathbb{Z}_6$	$3\mathbb{Z}_6$	$2\mathbb{Z}_6 + 3\mathbb{Z}_6$	$2\mathbb{Z}_6$	$3\mathbb{Z}_6$	$2\mathbb{Z}_6 + 3\mathbb{Z}_6$
0	0	0	0	0	0
0	0	0	0	3	3
4	0	4	0	0	0
4	0	4	0	3	3
2	0	2	0	0	0
2	0	2	0	3	3

We omit the tables for other scalars.

Quiz 9.10 *Write down the tables of* $4\times$ *and* $5\times$. *Show that the following two module homomorphisms are projections to the direct summands:*

$$3\times : 2\mathbb{Z}_6 \oplus 3\mathbb{Z}_6 \to 3\mathbb{Z}_6,$$
$$4\times : 2\mathbb{Z}_6 \oplus 3\mathbb{Z}_6 \to 2\mathbb{Z}_6.$$

Example 9.39 Let $A = C^\infty(\mathbb{R}, \mathbb{R})$ be the set of infinitely differentiable functions $\mathbb{R} \to \mathbb{R}$. We define addition $f + g$ and multiplication fg for $f, g \in A$ pointwise as follows:

$$f + g : x \mapsto f(x) + g(x),$$
$$fg : x \mapsto f(x)g(x).$$

Then, A is a commutative ring with the zero and unity being

$$0_A : x \mapsto 0 \quad \text{and} \quad 1_A : x \mapsto 1,$$

respectively.

The function $x \mapsto x$ does not have an inverse for multiplication since $x \mapsto \dfrac{1}{x}$ does not make sense at $x = 0$. Thus, A is not a field.

The ring A is itself an A-module. Consider an A-submodule U defined by

$$U = \{f \in A \mid f(0) = 0\}.$$

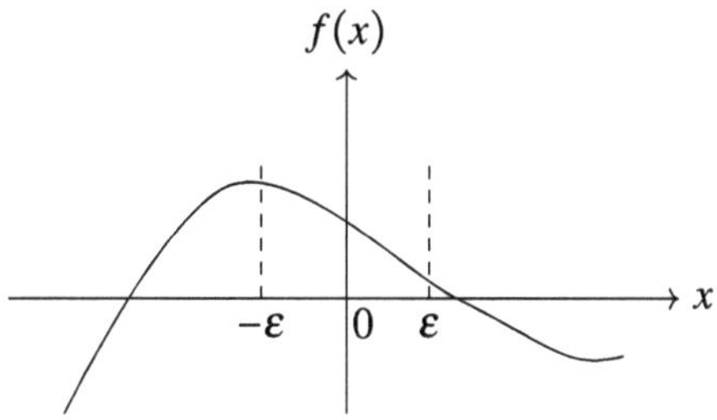

Fig. 9.3 $f(x) > 0$ holds for $-\varepsilon < x < \varepsilon$

We can easily verify that U is closed under A-scalar multiplication. We wonder if U has a complement. Suppose V is a complementary A-submodule of U. There must be $f \in V$ such that $f(0) \neq 0$; otherwise we have $V \subset U$, which contradicts the assumption. We can assume $f(0) > 0$ without loss of generality. Then, there is a positive real number ε that guarantees

$$f(x) > 0 \quad \text{for} \quad -\varepsilon < x < \varepsilon,$$

which is the consequence of continuity. See Fig. 9.3. Let $g(x) = xf(x)$, which must be in V, and $g(0) = 0$. Then, we have

$$g \in U \cap V \quad \text{and} \quad g \neq 0_A.$$

Thus, V cannot be a complement of U.

Example 9.40 The set $A = \left\{ \begin{pmatrix} a & b \\ 0 & c \end{pmatrix} \middle| a, b, c \in \mathbb{C} \right\}$ is a ring. We regard A as a left A-module. Here are some left A-submodules of A:

$$L_0 = \left\{ \begin{pmatrix} a & 0 \\ 0 & 0 \end{pmatrix} \middle| a \in \mathbb{C} \right\},$$

$$L_\infty = \left\{ \begin{pmatrix} 0 & b \\ 0 & c \end{pmatrix} \middle| b, c \in \mathbb{C} \right\},$$

$$N = \left\{ \begin{pmatrix} 0 & b \\ 0 & 0 \end{pmatrix} \middle| b \in \mathbb{C} \right\}.$$

The submodules L_0 and L_∞ are complementary submodules. The submodule N has no complement.

Quiz 9.11 *Prove that L_0, L_∞, and N are left A-modules.*

Exercises

[1] Consider column vectors $\begin{pmatrix} a \\ b \end{pmatrix} \in \mathbb{R}^2$. Let the left scalar multiplication $\bullet$ by $x + y\sqrt{-1} \in \mathbb{C}$ be defined as

$$(x + y\sqrt{-1}) \bullet \begin{pmatrix} a \\ b \end{pmatrix} = \begin{pmatrix} x & -y \\ y & x \end{pmatrix} \begin{pmatrix} a \\ b \end{pmatrix} = \begin{pmatrix} xa - yb \\ ya + xb \end{pmatrix}.$$

(1) For any $x + y\sqrt{-1}, u + v\sqrt{-1} \in \mathbb{C}$ $(x, y, u, v \in \mathbb{R})$, show that

$$((u+v\sqrt{-1})(x+y\sqrt{-1})) \bullet \begin{pmatrix} a \\ b \end{pmatrix} = (u+v\sqrt{-1}) \bullet \left((x + y\sqrt{-1}) \bullet \begin{pmatrix} a \\ b \end{pmatrix} \right).$$

(2) Prove the following equation:

$$((x+y\sqrt{-1})+(u+v\sqrt{-1})) \bullet \begin{pmatrix} a \\ b \end{pmatrix} = (x+y\sqrt{-1}) \bullet \begin{pmatrix} a \\ b \end{pmatrix} + (u+v\sqrt{-1}) \bullet \begin{pmatrix} a \\ b \end{pmatrix}.$$

(3) Prove the following equation:

$$(x+y\sqrt{-1}) \bullet \left(\begin{pmatrix} a \\ b \end{pmatrix} + \begin{pmatrix} c \\ d \end{pmatrix} \right) = (x+y\sqrt{-1}) \bullet \begin{pmatrix} a \\ b \end{pmatrix} + (x+y\sqrt{-1}) \bullet \begin{pmatrix} c \\ d \end{pmatrix}.$$

(4) Prove the following equations:

$$1 \bullet \begin{pmatrix} a \\ b \end{pmatrix} = \begin{pmatrix} a \\ b \end{pmatrix}, \quad 0 \bullet \begin{pmatrix} a \\ b \end{pmatrix} = \begin{pmatrix} 0 \\ 0 \end{pmatrix}.$$

In this way, we can introduce a left $\mathbb{C}$-module structure to $\mathbb{R}^2$.

[2] Consider column vectors $\begin{pmatrix} a \\ b \end{pmatrix} \in \mathbb{R}^2$. Let the left scalar multiplication $\odot$ by $x + y\sqrt{-1} \in \mathbb{C}$ be defined as

$$(x + y\sqrt{-1}) \odot \begin{pmatrix} a \\ b \end{pmatrix} = \begin{pmatrix} x + y & -2y \\ y & x - y \end{pmatrix} \begin{pmatrix} a \\ b \end{pmatrix} = \begin{pmatrix} (x + y)a - 2yb \\ ya + (x - y)b \end{pmatrix}.$$

Repeat the similar calculations in (1)–(4) of Exercise [1] above. In this way, we can introduce a different left $\mathbb{C}$-module structure to $\mathbb{R}^2$. At this point, most readers realize that there are infinitely many $\mathbb{C}$-module structures for $\mathbb{R}^2$. Further, $\mathbb{C}$-module structures can be continuously deformed. Let P be any real 2×2 regular matrix. We can define scalar multiplication $\bullet_P$ as

$$(x + y\sqrt{-1}) \bullet_P \begin{pmatrix} a \\ b \end{pmatrix} = P \begin{pmatrix} x & -y \\ y & x \end{pmatrix} P^{-1} \begin{pmatrix} a \\ b \end{pmatrix}.$$

[3] Consider the ring A in Example 9.32. Define a $\mathbb{C}$-linear map $\varphi : \mathbb{C}^3 \to \mathbb{C}^3$ by

$$\varphi : \begin{pmatrix} x \\ y \\ z \end{pmatrix} \mapsto \begin{pmatrix} 1 & 3 & 0 \\ 0 & 2 & 0 \\ 0 & 0 & 1 \end{pmatrix} \begin{pmatrix} x \\ y \\ z \end{pmatrix}.$$

Show that φ is not a left A-module homomorphism. It is sufficient to find a scalar that does not commute with φ.

[4] Verify the explicit descriptions of $\ker \varphi$ and $\operatorname{Im} \varphi$ in Example 9.33.

[5] Revisit Example 9.33. Let k be any field. Consider the polynomial ring $k[X]$. The field k is regarded as a $k[X]$-module. We define the scalar multiplication $f(X) \bullet z$ between the scalar $f(X)$ and a module element $z \in k$ by $f(1)z$.

We define a $k[X]$-module homomorphism $\varphi : k[X] \to k$ by

$$\varphi : f(X) \mapsto f(1).$$

Find $\ker \varphi$ and $\operatorname{Im} \varphi$.

[6] We consider a $\mathbb{Z}$-module $\mathbb{Z}$. Prove that non-trivial submodules of $\mathbb{Z}$ do not have complements.

[7] Revisit Example 9.40. For any $k \in \mathbb{C}$, we consider the following sets:

$$L_k = \left\{ \begin{pmatrix} a & ka \\ 0 & 0 \end{pmatrix} \middle| a \in \mathbb{C} \right\}, \; L_\infty = \left\{ \begin{pmatrix} 0 & b \\ 0 & c \end{pmatrix} \middle| b, c \in \mathbb{C} \right\}, \text{ and}$$

$$N = \left\{ \begin{pmatrix} 0 & b \\ 0 & 0 \end{pmatrix} \middle| b \in \mathbb{C} \right\}.$$

In the following, we assume p, q and r are non-zero complex numbers.

(1) Show that L_k and L_∞ are left A-submodules of A.
(2) Find complementary submodules of L_k and L_∞.
(3) Verify the following equations:

$$A \begin{pmatrix} p & 0 \\ 0 & 0 \end{pmatrix} = L_0, \quad A \begin{pmatrix} 0 & q \\ 0 & r \end{pmatrix} = L_\infty,$$

$$A \begin{pmatrix} p & q \\ 0 & r \end{pmatrix} = A, \quad A \begin{pmatrix} p & q \\ 0 & 0 \end{pmatrix} = L_{q/p},$$

$$A \begin{pmatrix} p & 0 \\ 0 & r \end{pmatrix} = A, \quad A \begin{pmatrix} 0 & 0 \\ 0 & r \end{pmatrix} = L_\infty,$$

$$A \begin{pmatrix} 0 & q \\ 0 & 0 \end{pmatrix} = N.$$

(4) Show that N has no complement.

Chapter 10
Ideals

A ring A is a left or right A-module or A-bimodule. We study A-module structure of A itself. We introduce the notions of ideals and factor rings/modules, which correspond to normal subgroups and factor groups in group theory.

10.1 Ideals and Factor Rings

An ideal is a submodule of a ring. Ideals are kernels of ring homomorphisms, just as normal subgroups are kernels of group homomorphisms.

Definition 10.1 When a ring A is considered a left A-module, a left A-submodule of A is called a *left ideal* of A. When a ring A is considered a right A-module, a right A-submodule of A is called a *right ideal* of A. When a ring A is considered a A-bimodule, a submodule of A is called a *two-sided ideal* of A or sometimes simply an *ideal* of A. If A is a commutative ring, an A-submodule of A is simply called an *ideal*.

Example 10.2 We consider the ring A:

$$A = \left\{ \begin{pmatrix} a_{11} & a_{12} & 0 \\ a_{21} & a_{22} & 0 \\ 0 & 0 & a_{33} \end{pmatrix} \middle| a_{ij} \in \mathbb{C} \right\}.$$

Let us consider two additive subgroups I_1 and I_2 of A:

$$I_1 = \left\{ \begin{pmatrix} x_{11} & x_{12} & 0 \\ x_{21} & x_{22} & 0 \\ 0 & 0 & 0 \end{pmatrix} \middle| x_{ij} \in \mathbb{C} \right\}, \quad I_2 = \left\{ \begin{pmatrix} 0 & 0 & 0 \\ 0 & 0 & 0 \\ 0 & 0 & x_{33} \end{pmatrix} \middle| x_{33} \in \mathbb{C} \right\}.$$

S. Yukita, *First Lectures in Algebra*, Springer Asia Pacific Mathematics Series 10,
https://doi.org/10.1007/978-981-95-5457-7_10

Both are two-sided ideals of A.

Let us consider next two additive subgroups I_3 and I_4 of A:

$$I_3 = \left\{ \begin{pmatrix} x_{11} & 0 & 0 \\ x_{21} & 0 & 0 \\ 0 & 0 & 0 \end{pmatrix} \middle| x_{ij} \in \mathbb{C} \right\}, \quad I_4 = \left\{ \begin{pmatrix} x_{11} & x_{12} & 0 \\ 0 & 0 & 0 \\ 0 & 0 & 0 \end{pmatrix} \middle| x_{ij} \in \mathbb{C} \right\}.$$

The additive subgroup I_3 is a left ideal but not a right ideal of A. The additive subgroup I_4 is a right ideal but not a left ideal of A.

Quiz 10.1 Show that I_1 and I_2 in Example 10.2 are closed under left and right scalar multiplications.

Quiz 10.2 Show that I_3 in Examples 10.2 is closed under the left scalar multiplication but not under the right one. Show that I_4 is closed under the right scalar multiplication but not under the left one.

Recall that the kernels of group homomorphisms are normal subgroups. In ring theory, ideals play the role of normal subgroups.

Theorem 10.3 *Let $\varphi : A \to B$ be a ring homomorphism. Then,* $\ker \varphi$ *is a two-sided ideal of A.*

Proof Since φ is a group homomorphism of additive groups, it remains to prove that φ is closed under the left and right scalar multiplications. For any $a \in A$ and $x \in \ker \varphi$ we have

$$\varphi(ax) = \varphi(a)\varphi(x) = \varphi(a)0 = 0 \quad \text{and} \quad \varphi(xa) = \varphi(x)\varphi(a) = 0\varphi(a) = 0,$$

which shows that $\ker \varphi$ is closed under the scalar multiplication from both sides. □

Given a ring homomorphism $\varphi : A \to B$, the ring B can be regarded as a left A-module. The scalar multiplication is defined by

$$a \bullet x = \varphi(a)x$$

for $a \in A$ and $x \in B$. Likewise, B can be regarded as a right A-module.

Given a ring homomorphism $\varphi : A \to B$, as stated above, the ring B is a left A-module. The additive subgroup $\operatorname{Im} \varphi$ is a left A-submodule of B. We have the following isomorphism of left A-modules:

$$A/\ker \varphi \simeq \operatorname{Im} \varphi.$$

Suppose there is a two-sided ideal I of a ring A. The additive factor group A/I is a A-bimodule. We can regard A/I as a ring as follows. For any $a + I, b + I \in A/I$, addition is defined by

$$(a + I) + (b + I) = (a + b) + I,$$

which is well-defined as an additive factor group. We want to define multiplication $\bullet$ by

$$(a + I) \bullet (b + I) = ab + I.$$

We have to see if it is well-defined. Let a' and b' be alternative representatives for a and b, respectively. Namely, we have $a' - a \in I$ and $b' - b \in I$. Then, we have

$$\begin{aligned} & (a' + I) \bullet (b' + I) \\ = \; & (a + (a' - a) + I) \bullet (b + (b' - b) + I) \\ = \; & (a + I) \bullet (b + I), \end{aligned}$$

which means the result is independent of the choice of representatives. From now on, we omit the symbol $\bullet$. Next, we verify that the distributive law holds. We have to show

$$((a + I) + (b + I))(c + I) = (a + I)(c + I) + (b + I)(c + I).$$

The LHS is

$$((a + I) + (b + I))(c + I) = (a + b + I)(c + I) = (a + b)c + I = ac + bc + I.$$

The RHS is

$$(a + I)(c + I) + (b + I)(c + I) = (ac + I) + (bc + I) = ac + bc + I.$$

Thus, both sides coincide. Likewise, we can show

$$(c + 1)((a + I) + (b + I)) = (c + I)(a + I) + (c + I)(b + I).$$

The zero element is $0 + I$. The unity is $1 + I$. Thus, we established that A/I is a ring.

Definition 10.4 Let I be a two-sided ideal of a ring A. The additive factor group A/I can be regarded as a ring. The ring A/I is called a *factor ring*, *quotient ring*, or *residue class ring*.

Example 10.5 We define a ring homomorphism $\varphi : \mathbb{C}[X] \to \mathbb{C}$ by

$$\varphi : f(X) \mapsto f(1).$$

The map φ takes a polynomial and returns a value obtained by substituting X with 1. Addition and multiplication commute with the substitution operation. This means φ is a ring homomorphism.

We regard $\mathbb{C}$ as a $\mathbb{C}[X]$-module with the scalar multiplication $f(X)\bullet$ defined by

$$f(X) \bullet z = f(1)z$$

for any $f(X) \in \mathbb{C}[X]$ and $z \in \mathbb{C}$. We have

$$\varphi(f(X) \bullet z) = f(1)z \quad \text{and} \quad f(X) \bullet \varphi(z) = f(1)z,$$

where z in $\varphi(z)$ is considered a polynomial with constant term only, namely $\varphi(z) = z$. Thus, we have $\varphi(f(X) \bullet z) = f(X) \bullet \varphi(z)$, which means that φ is a $\mathbb{C}[X]$-module homomorphism. The submodule $\ker\varphi$ consists of all the polynomials with $f(1) = 0$. The factor theorem says such polynomials are just those that have $X - 1$ as a factor:

$$\ker\varphi = (X - 1)\mathbb{C}[X] = \{(X - 1)f(X) \mid f(X) \in \mathbb{C}[X]\}.$$

We have the following $\mathbb{C}[X]$-module isomorphism:

$$\mathbb{C} \simeq \mathbb{C}[X]/\ker\varphi = \mathbb{C}[X]/((X - 1)\mathbb{C}[X]),$$

which is a ring isomorphism at the same time.

Example 10.6 We define a ring homomorphism $\varphi : \mathbb{C}[X] \to \mathbb{C}$ by

$$\varphi : f(X) \mapsto f(2).$$

The map φ takes a polynomial and returns a value obtained by substituting X with 2. Like Example 10.5, φ is a ring homomorphism. We regard $\mathbb{C}$ as a $\mathbb{C}[X]$-module with the scalar multiplication $f(X)\bullet$ defined by

$$f(X) \bullet z = f(2)z$$

for any $f(X) \in \mathbb{C}[X]$ and $z \in \mathbb{C}$. We have the following $\mathbb{C}[X]$-module isomorphism:

$$\mathbb{C} \simeq \mathbb{C}[X]/\ker\varphi = \mathbb{C}[X]/((X - 2)\mathbb{C}[X]),$$

which is a ring isomorphism at the same time.

There are infinitely many ways of regarding $\mathbb{C}$ as a $\mathbb{C}[X]$-module as this and previous examples suggest.

Example 10.7 Let $\varphi : \mathbb{C} \to \mathbb{C}[X]$ be the map that sends each complex number z to the polynomial that has only constant term z. The map φ is a ring homomorphism.

The following two theorems concern left ideals. Recall that if A is a commutative ring, then we can simply say ideals.

Theorem 10.8 *Let I and J be left (right, or two-sided) ideals of a ring A. Then, the intersection $I \cap J$ is a left (right, or two-sided) ideal.*

Proof We have already known that $I \cap J$ is an additive subgroup of A. It remains to prove that $I \cap J$ is closed under the left scalar multiplication. For any $a \in A$ and $x \in I \cap J$, we have:

$$\begin{aligned} x \in I \quad &\text{therefore} \quad ax \in I, \quad \text{and} \\ x \in J \quad &\text{therefore} \quad ax \in J, \end{aligned}$$

which gives $ax \in I \cap J$.

Right and two-sided ideals can be treated in the same way. □

Theorem 10.9 *Let I and J be left (right, or two-sided) ideals of a ring A. Then, the set*

$$I + J = \{x + y \mid x \in I, y \in J\}$$

is a left (right, or two-sided) ideal.

Proof We have already known that $I + J$ is an additive subgroup of A. It remains to prove that $I + J$ is closed under the left scalar multiplication. For any $a \in A$, $x \in I$ and $y \in J$, we can distribute as

$$a(x + y) = ax + ay.$$

Since $ax \in I$ and $ay \in J$, we have $ax+ay \in I+J$. Thus, we have $a(x+y) \in I+J$.

Right and two-sided ideals can be treated in the same way. □

Example 10.10 Ideals $2\mathbb{Z}, 3\mathbb{Z}, 4\mathbb{Z}$, and $6\mathbb{Z}$ of the ring $\mathbb{Z}$ have the following equations:

$$\begin{aligned} 2\mathbb{Z} \cap 3\mathbb{Z} &= 6\mathbb{Z}, \\ 2\mathbb{Z} + 3\mathbb{Z} &= \mathbb{Z}, \\ 4\mathbb{Z} \cap 6\mathbb{Z} &= 12\mathbb{Z}, \\ 4\mathbb{Z} + 6\mathbb{Z} &= 2\mathbb{Z}. \end{aligned}$$

Quiz 10.3 Conjecture any laws from the example above.

Example 10.11 We consider ideals $X\,\mathbb{C}[X, Y]$ and $Y\,\mathbb{C}[X, Y]$ of the polynomial ring $\mathbb{C}[X, Y]$. We have

$$X\,\mathbb{C}[X, Y] \cap Y\,\mathbb{C}[X, Y] = XY\,\mathbb{C}[X, Y].$$

We prove this.

Suppose $f(X, Y) \in X\,\mathbb{C}[X, Y] \cap Y\,\mathbb{C}[X, Y]$. Since $f(X, Y) \in X\,\mathbb{C}[X, Y]$, the polynomial $f(X, Y)$ has X as a factor. Since $f(X, Y) \in Y\,\mathbb{C}[X, Y]$, the polynomial $f(X, Y)$ has Y as a factor. So, XY is a factor of $f(X, Y)$. Namely, we have

$$X\,\mathbb{C}[X, Y] \cap Y\,\mathbb{C}[X, Y] \subset XY\,\mathbb{C}[X, Y].$$

Conversely, suppose $f(X, Y) \in XY\,\mathbb{C}[X, Y]$. Since $XY\,\mathbb{C}[X, Y] \subset X\,\mathbb{C}[X, Y]$ and $XY\,\mathbb{C}[X, Y] \subset Y\,\mathbb{C}[X, Y]$, we have

$$XY\,\mathbb{C}[X, Y] \subset X\,\mathbb{C}[X, Y] \cap Y\,\mathbb{C}[X, Y].$$

Example 10.12 We consider again ideals $X\,\mathbb{C}[X, Y]$ and $Y\,\mathbb{C}[X, Y]$ of the polynomial ring $\mathbb{C}[X, Y]$. We have

$$X\,\mathbb{C}[X, Y] + Y\,\mathbb{C}[X, Y] = \{f(X, Y) \in \mathbb{C}[X, Y] \mid f(0, 0) = 0\}.$$

This is a typical example of a maximal ideal. We prove this equality.

Set $I = X\,\mathbb{C}[X, Y] + Y\,\mathbb{C}[X, Y]$. Any element $f(X, Y) \in I$ must be expressed as

$$f(X, Y) = X\,g(X, Y) + Y\,h(X, Y).$$

Then, we have $f(0, 0) = 0$.

Conversely, take any $f(X, Y) \in \mathbb{C}[X, Y]$ such that $f(0, 0) = 0$. The Taylor expansion of $f(X, Y)$ at $(0, 0)$ is

$$f(X, Y) = c_0 + c_{10}X + c_{01}Y + c_{20}X^2 + c_{11}XY + c_{02}Y^2 + \cdots \quad \text{(finite sum)}.$$

$f(0, 0) = 0$ means $c_0 = 0$. Collecting all terms without X in $f(X, Y)$, we have $Y\,h(Y)$ for some $h(Y) \in \mathbb{C}[X, Y]$. Since all the terms of polynomial $f(X, Y) - Y\,h(Y)$ contain X, it can be expressed as $X\,g(X, Y)$ for some $g(X, Y) \in \mathbb{C}[X, Y]$. Thus, we have

$$f(X, Y) = X\,g(X, Y) + Y\,h(Y) \in I.$$

If we define a ring homomorphism $\varphi : \mathbb{C}[X, Y] \to \mathbb{C}$ by

$$\varphi : f(X, Y) \mapsto f(0, 0),$$

then we have

$$\ker \varphi = X\,\mathbb{C}[X, Y] + Y\,\mathbb{C}[X, Y].$$

10.2 Ideals of Polynomial Rings

In this section, we study ideals of the rings $\mathbb{C}[X]$, $\mathbb{R}[X]$, and $\mathbb{Q}[X]$. We need not distinguish left and right ideals.

Example 10.13 We consider an ideal $I = X\,\mathbb{C}[X]$ of the ring $\mathbb{C}[X]$. The factor ring is the following:

$$\mathbb{C}[X]/I \simeq \mathbb{C},$$

where a ring isomorphism $\varphi : \mathbb{C}[X]/I \to \mathbb{C}$ is given by

$$\varphi : f(X) \mapsto f(0).$$

Example 10.14 We consider an ideal $I = X^3\,\mathbb{C}[X]$ of the ring $\mathbb{C}[X]$. The factor ring is the following:

$$\mathbb{C}[X]/I = \{(aX^2 + bX + c) + I \mid a, b, c \in \mathbb{C}\}.$$

In fact, Euclid's division lemma guarantees that any $f(X) \in \mathbb{C}[X]$ can be expressed uniquely as

$$f(X) = X^3\,g(X) + aX^2 + bX + c,$$

where $g(X)$ is the quotient and aX^2+bX+c is the remainder. We express an I-orbit (coset, residue class) as

$$(aX^2 + bX + c) + I = \overline{aX^2 + bX + c}.$$

The addition in this factor ring $\mathbb{C}[X]/I$ is straightforward:

$$\overline{aX^2 + bX + c} + \overline{dX^2 + eX + f} = \overline{(a+d)X^2 + (b+e)X + c + f}.$$

The multiplication is given by the following formula:

$$\overline{aX^2 + bX + c} \cdot \overline{dX^2 + eX + f} = \overline{(af + be + cd)X^2 + (bf + ce)X + cf}.$$

Quiz 10.4 Show that $\overline{X} \cdot \overline{X^2} = \bar{0}$ in the above example.

There are cases in which multiplication between non-zero elements yields zero.

Definition 10.15 If there is a non-zero element a and b of a ring A such that $ab = 0$, elements a and b are called a *left* and *right zero divisor*, respectively. If A is a commutative ring, then they are simply called *zero divisors*.

Example 10.16 We immediately find zero divisors in matrix rings. Let

$$P_1 = \begin{pmatrix} 1 & 0 \\ 0 & 0 \end{pmatrix} \quad \text{and} \quad P_2 = \begin{pmatrix} 0 & 0 \\ 0 & 1 \end{pmatrix}$$

be elements of the ring $M_2(\mathbb{R})$. Then, the product $P_1 P_2$ is the zero matrix. The matrix P_1 is a left divisor and P_2 is a right divisor.

If we regard 2×2 matrix as a linear map from the column vector space $\mathbb{R}^2$ to itself, then P_1 and P_2 are projections to the x and y-axes, respectively.

Fields do not have zero divisors. Let $a, b \neq 0$ and $ab = 0$ in some field. There is an inverse a^{-1}. If we multiply both sides by a^{-1} from the left, we get $b = 0$, which contradicts the assumption.

Definition 10.17 A commutative ring is called an *integral domain* if it has no zero divisors.

Example 10.18 We give some examples of integral domains. Commutative fields are integral domains. The rings $\mathbb{Z}$, $\mathbb{C}[X]$, $\mathbb{R}[X]$, and $\mathbb{Q}[X]$ are integral domains.

The factor ring $A = \mathbb{C}[X]/(X^3\,\mathbb{C}[X])$ is not integral. In fact, we have

$$\overline{X}\,\overline{X^2} = \overline{X^3} = \bar{0}.$$

The ring $\mathbb{Z}/6\mathbb{Z} = \{\bar{0}, \bar{1}, \ldots, \bar{5}\}$ is not integral since $\bar{2} \times \bar{3} = \bar{0}$.

Let A be an integral domain. Given an ideal I, the factor ring A/I may or may not be integral.

Definition 10.19 Let P be a two-sided ideal of a ring A such that $P \subsetneqq A$. The ideal P is called a *prime ideal* if it has no zero divisors. In other words, P is called a *prime ideal* if it satisfies the following condition:

- If $a, b \in P$ and $ab \in P$, then $a \in P$ or $b \in P$.

Remark 10.20 If the ring A is commutative, an ideal P is a prime ideal if and only if the factor ring A/P is integral.

Example 10.21 Consider an ideal $I = (X^2 - 2)\,\mathbb{Q}[X]$ of the ring $\mathbb{Q}[X]$. Any $f(X) \in \mathbb{Q}[X]$ can be expressed as

$$f(X) = (X^2 - 2)Q(X) + aX + b.$$

The factor ring is

$$\mathbb{Q}[X]/I = \{\overline{aX + b} \mid a, b \in \mathbb{Q}\},$$

where $\overline{aX + b}$ is the I-orbit (residue class) passing $aX + b$. The addition formula is given by

$$\overline{aX+b}+\overline{cX+d}=\overline{(a+c)X+b+d}.$$

The multiplication formula is given by

$$\overline{aX+b}\times\overline{cX+d}=\overline{(ad+bc)X+bd+2ac}.$$

The former formula is straightforward. The latter one is obtained as follows. We calculate the product of representatives, then apply Euclidean division by X^2-2.

$$\begin{aligned}&(aX+b)\times(cX+d)\\=\ &acX^2+(ad+bc)X+bd\\=\ &ac(X^2-2)\times 1+(ad+bc)X+bd+2ac.\end{aligned}$$

The remainder $(ad+bc)X+bd+2ac$ is a unique representative since we can restrict representatives to polynomials of degree 1.

To show that the factor ring is a field, we seek a reciprocal (multiplicative inverse) formula for $\overline{aX+b}\neq\bar{0}$. Setting

$$\overline{aX+b}\cdot\overline{cX+d}=\overline{(ad+bc)X+bd+2ac}=\bar{1},$$

we consider the system of equations with unknowns c, d:

$$\begin{cases}bc+ad=0\\bd+2ac=1\end{cases}.$$

Putting it in a matrix form, we have

$$\begin{pmatrix}b & a\\2a & b\end{pmatrix}\begin{pmatrix}c\\d\end{pmatrix}=\begin{pmatrix}0\\1\end{pmatrix}.$$

If $\det\begin{pmatrix}b & a\\2a & b\end{pmatrix}=b^2-2a^2\neq 0$, then this equation has a unique solution.

Suppose $b^2-2a^2=0$. If $b=0$, then $a=0$. This contradicts $\overline{aX+b}\neq\bar{0}$. Thus, we have $b\neq 0$ and the following:

$$2=\left(\frac{a}{b}\right)^2\quad\text{therefore}\quad\frac{a}{b}=\sqrt{2}.$$

Since $a, b\in\mathbb{Q}$, this is impossible. Now, we have the formula for reciprocals:

$$\overline{aX+b}^{\,-1}=\overline{-\frac{a}{b^2-2a^2}X+\frac{b}{b^2-2a^2}}.$$

The factor ring $\mathbb{Q}[X]/((X^2-2)\,\mathbb{Q}[X])$ is a field, and consequently an integral domain. The ideal $(X^2-2)\,\mathbb{Q}[X]$ is prime.

Example 10.22 Consider an ideal $I_1 = X\,\mathbb{C}[X, Y]$ of the ring $\mathbb{C}[X, Y]$. We define a ring homomorphism $\varphi_1 : \mathbb{C}[X, Y] \to \mathbb{C}[Y]$ by

$$\varphi_1 : f(X, Y) \mapsto f(0, Y).$$

We have $\ker \varphi_1 = I_1$ and $\mathbb{C}[X, Y]/I_1 \simeq \mathbb{C}[Y]$. Since $\mathbb{C}[Y]$ is an integral domain, I_1 is a prime ideal of $\mathbb{C}[X, Y]$.

Consider an ideal $I_2 = X\,\mathbb{C}[X, Y] + Y\,\mathbb{C}[X, Y]$ of the ring $\mathbb{C}[X, Y]$. We define a ring homomorphism $\varphi_2 : \mathbb{C}[X, Y] \to \mathbb{C}$ by

$$\varphi_2 : f(X, Y) \mapsto f(0, 0).$$

We have $\ker \varphi_2 = I_2$ and $\mathbb{C}[X, Y]/I_2 \simeq \mathbb{C}$. See Example 10.12. Since $\mathbb{C}$ is a field and consequently integral domain, the ideal I_2 is prime.

We keep the following relations in mind for later study:

$$\{0\} \subset I_1 \subset I_2 \subset \mathbb{C}[X, Y].$$

10.3 Ideals of Factor Rings and Maximal Ideals

We have discussed subgroups of a factor group in Sect. 6.2. We can do similar arguments for factor rings. In this section, we limit our discussion to commutative rings.

Definition 10.23 An ideal I of the ring A is called a *maximal ideal* of A if it satisfies the following condition:

- There is no ideal J such that $I \subsetneqq J \subsetneqq A$.

Example 10.24 For any prime number p, the ideal $p\mathbb{Z}$ of the ring $\mathbb{Z}$ is a maximal ideal of $\mathbb{Z}$. See Fig. 10.1.

Factor rings of maximal ideals have a special property.

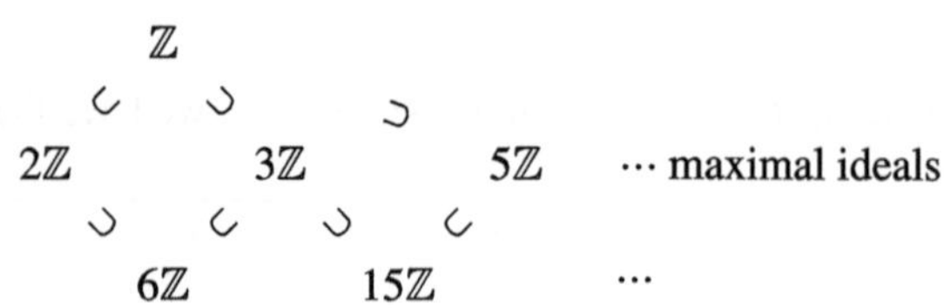

Fig. 10.1 Maximal ideals of $\mathbb{Z}$

Theorem 10.25 *Let I be an ideal of a ring A. The ideal I is maximal if and only if the factor ring A/I is a field.*

We may replace this with the definition of maximal ideals. To prove Theorem 10.25, we prepare some lemmas.

Lemma 10.26 *Let I be an ideal of a ring A. We consider the following two sets:*

$$\Gamma = \{K \mid K \text{ is an ideal of } A,\ K \supset I\},$$

$$\Delta = \{L \mid L \text{ is an ideal of } A/I\}.$$

We define a map $\varphi : \Gamma \to \Delta$ by

$$\varphi : K \mapsto K/I.$$

We have the following:

(1) The map φ is a bijection.
(2) The map φ preserves the inclusion relation.

Proof As additive groups, (1) and (2) have already been established by Theorem 6.14. It remains to prove that K/I is an ideal of the factor ring A/I for all $K \in \Gamma$. For any $a + I \in A/I$ and $k + I \in K/I$, we have

$$(a + I) \bullet (k + I) = ak + I.$$

Since K is an ideal of A, we have $ak \in K$. Thus, we have $ak + I \in K/I$. This shows that K/I is an ideal of the factor ring A/I. □

Lemma 10.27 *Let I be an ideal of a field F. Then, $I = \{0\}$ or $I = F$.*

Proof Let $I \neq \{0\}$ be an ideal of F. Then, there is an element $a \in I$ such that $a \neq 0$. Since F is a field, there is a^{-1}. We have

$$1 = a^{-1}a \in I,$$

which means $I = F$. □

Lemma 10.28 *If a ring F has only $\{0\}$ and F as its ideals, then F is a field.*

Proof Take any $a \in F\backslash\{0\}$. The set aF is an ideal of F. By assumption, $aF = \{0\}$ or $aF = F$. The former case leads to $a \bullet 1 = a = 0$, which contradicts $a \neq 0$. Only the latter is possible. Then, there is an element $b \in F$ such that $ab = 1$. We conclude that F is a field. □

Proof of Theorem 10.25 Suppose A/I is a field. Lemma 10.27 tells us that

$$\Delta = \{\{0\}, A/I\}.$$

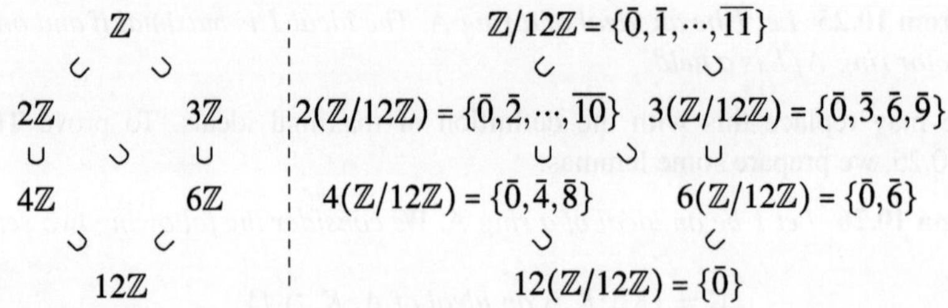

Fig. 10.2 Correspondence between Γ and Δ

Lemma 10.26 tells us that φ is a bijection. Therefore, we have

$$\Gamma = \{I, A\},$$

which means that I is a maximal ideal of A.

Conversely, suppose that I is a maximal ideal of A, namely

$$\Gamma = \{I, A\}.$$

By Lemma 10.26, we have

$$\Delta = \{\{0\}, A/I\}.$$

Lemma 10.28 tells us that A/I is a field. □

Remark 10.29 Maximal ideals are prime ideals. The converse is not true. We have already seen many examples.

Example 10.30 Consider an ideal $I = 12\mathbb{Z}$ of the ring $\mathbb{Z}$. The ideals of the factor ring $\mathbb{Z}/12\mathbb{Z} = \{\bar{0}, \bar{1}, \ldots, \overline{11}\}$ and the ideals that contain I correspond in a way displayed in Fig. 10.2.

10.4 Chinese Remainder Theorem

Consider the following system of congruence equations:

$$\begin{cases} x \equiv 2 \mod 3 \\ x \equiv 3 \mod 5 \\ x \equiv 2 \mod 7. \end{cases}$$

This is solved by the well-known *Chinese remainder theorem* or *Sunzi's theorem*.

Theorem 10.31 *Let natural numbers $m_1, m_2, \cdots, m_k$ be pairwise coprime. We have a $\mathbb{Z}$-module isomorphism:*

$$\mathbb{Z}/(m_1 m_2 \cdots m_k \mathbb{Z}) \simeq \mathbb{Z}/(m_1\mathbb{Z}) \oplus \mathbb{Z}/(m_2\mathbb{Z}) \oplus \cdots \oplus \mathbb{Z}/(m_k\mathbb{Z}).$$

Remark 10.32 In other words, the system of congruence equations

$$\begin{cases} x \equiv a_1 \mod m_1 \\ x \equiv a_2 \mod m_2 \\ \quad \vdots \\ x \equiv a_k \mod m_k \end{cases}$$

has a unique solution $x \mod m_1 m_2 \cdots m_k$.

To prove the above theorem we use the following lemma repeatedly.

Lemma 10.33 *Let natural numbers m and n be mutually coprime. We have a $\mathbb{Z}$-module isomorphism:*

$$\mathbb{Z}/(mn\mathbb{Z}) \simeq \mathbb{Z}/(m\mathbb{Z}) \oplus \mathbb{Z}/(n\mathbb{Z}).$$

Proof We define the map $\varphi : \mathbb{Z}/(mn\mathbb{Z}) \to \mathbb{Z}/(m\mathbb{Z}) \oplus \mathbb{Z}/(n\mathbb{Z})$ by

$$\varphi : x + mn\mathbb{Z} \mapsto (x + m\mathbb{Z}, x + n\mathbb{Z}),$$

which is well defined since it is independent of the choice of representatives. In fact, if we take any other representative x' from $x + mn\mathbb{Z}$, we have

$$x' - x \in mn\mathbb{Z} \subset m\mathbb{Z},$$
$$x' - x \in mn\mathbb{Z} \subset n\mathbb{Z}.$$

Therefore, we have

$$(x' + m\mathbb{Z}, x' + n\mathbb{Z}) = (x + m\mathbb{Z}, x + n\mathbb{Z}).$$

(1) φ is a $\mathbb{Z}$-module homomorphism.
To see this, take any elements $x + mn\mathbb{Z}$ and $y + mn\mathbb{Z}$ from $\mathbb{Z}/(mn\mathbb{Z})$. Trace the diagram constructed by φ and the addition operation $+$:

$$\begin{array}{ccc} x+mn\mathbb{Z} & \underset{\varphi}{\longmapsto} & (x+m\mathbb{Z}, x+n\mathbb{Z}) \\ y+mn\mathbb{Z} & \longmapsto & (y+m\mathbb{Z}, y+n\mathbb{Z}) \\ +\Big\downarrow & \circlearrowright & \Big\downarrow + \\ x+y+mn\mathbb{Z} & \underset{\varphi}{\longmapsto} & (x+y+m\mathbb{Z}, x+y+n\mathbb{Z}) \end{array}$$

The two paths starting from the upper left corner to lower right lead to the same result. This shows that φ is a group homomorphism.

Take any element $x + mn\mathbb{Z}$ from $\mathbb{Z}/(mn\mathbb{Z})$ and any scalar $z \in \mathbb{Z}$. Trace the diagram constructed by φ and the scalar multiplication $z\bullet$:

$$\begin{array}{ccc} x+mn\mathbb{Z} & \overset{\varphi}{\longmapsto} & (x+m\mathbb{Z}, x+n\mathbb{Z}) \\ z\bullet\Big\downarrow & \circlearrowright & \Big\downarrow z\bullet \\ zx+mn\mathbb{Z} & \underset{\varphi}{\longmapsto} & (zx+m\mathbb{Z}, zx+n\mathbb{Z}) \end{array}$$

The two paths starting from the upper left corner to lower right lead to the same result. This shows that φ and scalar multiplication commute.

Now, we have established that φ is a $\mathbb{Z}$-module homomorphism.

(2) φ is injective.
To see this, assume

$$\varphi(x + mn\mathbb{Z}) = (x + m\mathbb{Z}, x + n\mathbb{Z}) = (\bar{0}, \bar{0}) \in \mathbb{Z}/(m\mathbb{Z}) \oplus \mathbb{Z}/(n\mathbb{Z}).$$

Then, we have $x \in m\mathbb{Z}$, $x \in n\mathbb{Z}$, and consequently $x \in m\mathbb{Z} \cap n\mathbb{Z} = mn\mathbb{Z}$, recalling that m and n are mutually coprime. Thus, we have $x + mn\mathbb{Z} = \bar{0} \in \mathbb{Z}/(mn\mathbb{Z})$.

(3) φ is surjective.
To see this, take any element $(u+m\mathbb{Z}, v+n\mathbb{Z})$ from $\mathbb{Z}/(m\mathbb{Z})\oplus\mathbb{Z}/(n\mathbb{Z})$. We seek an element x such that $\varphi(x) = (u + m\mathbb{Z}, v + n\mathbb{Z})$. Since m and n are mutually coprime, Corollary 1.8 tells us that there are integers a and b that satisfy

$$am + bn = 1. \tag{10.1}$$

Using these, set

$$x = bnu + amv.$$

Then, we have

$$\varphi(x) = (u + m\mathbb{Z}, v + n\mathbb{Z}). \tag{10.2}$$

In fact, we see

$$bn \equiv 1 \mod m \quad \text{and} \quad am \equiv 1 \mod n$$

from Eq. (10.1). Therefore, we have

$$x \equiv u \mod m \quad \text{and} \quad x \equiv v \mod n,$$

which means Eq. (10.2) holds.

To sum up, the map φ is a $\mathbb{Z}$-module isomorphism. □

Let us apply the method in the above proof to a concrete example.

Example 10.34 Let us consider the system of congruence equations:

$$\begin{cases} x \equiv 5 \mod 15 \\ x \equiv 3 \mod 8. \end{cases}$$

We immediately notice that 15 and 8 are mutually coprime. The Euclidean algorithm in Example 1.7 or 1.9 gives us the following sequence of steps:

$$\begin{pmatrix} 15 & 8 \end{pmatrix} \begin{pmatrix} 0 & 1 \\ 1 & -1 \end{pmatrix} = \begin{pmatrix} 8 & 7 \end{pmatrix},$$

$$\begin{pmatrix} 8 & 7 \end{pmatrix} \begin{pmatrix} 0 & 1 \\ 1 & -1 \end{pmatrix} = \begin{pmatrix} 7 & 1 \end{pmatrix},$$

$$\begin{pmatrix} 7 & 1 \end{pmatrix} \begin{pmatrix} 0 & 1 \\ 1 & -7 \end{pmatrix} = \begin{pmatrix} 1 & 0 \end{pmatrix}.$$

Concatenating these, we have

$$\begin{pmatrix} 15 & 8 \end{pmatrix} \begin{pmatrix} -1 & 8 \\ 2 & -15 \end{pmatrix} = \begin{pmatrix} 1 & 0 \end{pmatrix},$$

which shows

$$-1 \times 15 + 2 \times 8 = 1.$$

If we set

$$x = 5 \times (2 \times 8) + 3 \times (-1 \times 15) = 35,$$

we get a special solution of the given system. Lemma 10.33 tells us that all the solutions are given by integers x that satisfy $x \equiv 35 \mod 120$.

Now, we proceed to the general case.

Proof of Theorem 10.31 We repeatedly apply Lemma 10.33.

The uniqueness of prime factorization tells us that m_1 and $m_2 m_3 \cdots m_k$ are mutually coprime. Lemma 10.33 tells us that the following holds:

$$\mathbb{Z}/(m_1 m_2 \cdots m_k \mathbb{Z}) \simeq \mathbb{Z}/(m_1 \mathbb{Z}) \oplus \mathbb{Z}/(m_2 m_3 \cdots m_k \mathbb{Z}).$$

We apply the lemma again to $\mathbb{Z}/(m_2 m_3 \cdots m_k \mathbb{Z})$ and so on. □

We introduce an alternative proof by Gauss, which is highly symmetric and gives a practical algorithm.

Proof (By Gauss) We define the map

$$\varphi : \mathbb{Z}/(m_1 m_2 \cdots m_k \mathbb{Z}) \to \mathbb{Z}/(m_1 \mathbb{Z}) \oplus \mathbb{Z}/(m_2 \mathbb{Z}) \oplus \cdots \oplus \mathbb{Z}/(m_k \mathbb{Z})$$

by

$$\varphi : x + M\mathbb{Z} \mapsto (x + m_1 \mathbb{Z}, x + m_2 \mathbb{Z}, \ldots, x + m_k \mathbb{Z}),$$

where we set $M = m_1 m_2 \cdots m_k$. This is well-defined since it is independent of the choice of representatives. In fact, if we take any other representative x' from $x + M\mathbb{Z}$, we have $x' - x \in M\mathbb{Z} \subset m_i \mathbb{Z}$ for all i. Thus, we have

$$x' + m_i \mathbb{Z} = x + m_i \mathbb{Z} \quad \text{for all } i.$$

(1) φ is a $\mathbb{Z}$-module homomorphism.
To see this, follow the same argument in the proof of Lemma 10.33.

(2) φ is injective.
To see this, assume $x + m_i \mathbb{Z} = 0 + m_i \mathbb{Z}$, namely $x \in m_i \mathbb{Z}$, for all i. We have

$$x \in m_1 \mathbb{Z} \cap m_2 \mathbb{Z} \cap \cdots \cap m_k \mathbb{Z} = m_1 m_2 \cdots m_k \mathbb{Z},$$

where the equality comes from the assumption that m_i's are pairwise coprime. Thus, we have $x + M\mathbb{Z} = 0 + M\mathbb{Z}$.

(3) φ is surjective.
To show this, set $M_i = m_1 m_2 \cdots \widehat{m_i} \cdots m_k \quad (i = 1, \ldots, k)$, where $\widehat{m_i}$ means m_i is excluded in the product. Or in other words, $m_i M_i = M$.
Since m_i and M_i are mutually coprime, there are integers s_i and t_i that satisfy

$$s_i m_i + t_i M_i = 1.$$

Then, we have

$$t_i M_i \equiv 1 \mod m_i.$$

For any

$$(a_1 + m_1\mathbb{Z}, a_2 + m_2\mathbb{Z}, \ldots, a_k + m_k\mathbb{Z}),$$

set

$$x = a_1(t_1 M_1) + a_2(t_2 M_2) + \cdots + a_k(t_k M_k).$$

Since

$$\begin{cases} M_i \equiv 0 \mod m_j & (i \neq j) \\ t_i M_i \equiv 1 \mod m_i \end{cases},$$

we have

$$x \equiv a_i \mod m_i$$

for all i's, which means

$$\varphi(x + M\mathbb{Z}) = (a_1 + m_1\mathbb{Z}, a_2 + m_2\mathbb{Z}, \ldots, a_k + m_k\mathbb{Z}).$$

To sum up, the map φ is a $\mathbb{Z}$-module isomorphism. □

Example 10.35 Consider the system of congruence equations:

$$\begin{cases} x \equiv 1 & \mod 2 \\ x \equiv 2 & \mod 3 \\ x \equiv 4 & \mod 5 \end{cases}. \tag{10.3}$$

Set

$$m_1 = 2, \quad m_2 = 3, \quad m_3 = 5,$$
$$M_1 = 3 \times 5, \quad M_2 = 2 \times 5, \quad M_3 = 2 \times 3.$$

The Euclidean algorithm gives:

$$8 \times m_1 + (-1) \times M_1 = 1,$$
$$-3 \times m_2 + 1 \times M_2 = 1,$$

$$-1 \times m_3 + 1 \times M_3 = 1.$$

Then, we get a special solution of Eq. (10.3):

$$x = 1 \times (-1) \times M_1 + 2 \times 1 \times M_2 + 4 \times 1 \times M_3 = 29.$$

All the solutions are given by integers x that satisfy $x \equiv 29 \mod 30$.

Exercises

[1] Let us express the factor ring $\mathbb{Z}/n\mathbb{Z}$ as $\mathbb{Z}_n$. Regard $\mathbb{Z}_{12}$ as a $\mathbb{Z}_{12}$-module. Consider submodules of $\mathbb{Z}_{12}$.

(1) Find a submodule complementary to the submodule $4\mathbb{Z}_{12}$.
(2) Show that there is no submodule complementary to $2\mathbb{Z}_{12}$.

[2] Let us define a ring homomorphism $\varphi : \mathbb{C}[X, Y] \to \mathbb{C}$ by

$$\varphi : f(X, Y) \mapsto f(1, 2).$$

Since the substitution $X = 1, Y = 2$ and the ring operations commute, the map φ is a ring homomorphism. Show that

$$\ker \varphi = (X - 1)\,\mathbb{C}[X, Y] + (Y - 2)\,\mathbb{C}[X, Y].$$

[3] Consider an ideal $I = (X^2 + X + 1)\,\mathbb{C}[X]$ of the ring $\mathbb{C}[X]$. Find the multiplication formula of the factor ring $\mathbb{C}[X]/I$. Determine if the factor ring is a field.

[4] Consider an ideal $I = (X^2 + X + 1)\,\mathbb{Q}[X]$ of the ring $\mathbb{Q}[X]$. Find the multiplication formula of the factor ring $\mathbb{Q}[X]/I$. Determine if the factor ring is a field.

[5] Consider ideals of the ring $\mathbb{C}[X, Y, Z]$:

$$\begin{aligned} I_1 &= X\,\mathbb{C}[X, Y, Z], \\ I_2 &= X\,\mathbb{C}[X, Y, Z] + Y\,\mathbb{C}[X, Y, Z], \\ I_3 &= X\,\mathbb{C}[X, Y, Z] + Y\,\mathbb{C}[X, Y, Z] + Z\,\mathbb{C}[X, Y, Z]. \end{aligned}$$

(1) Show that $\mathbb{C}[X, Y, Z]/I_1 \simeq \mathbb{C}[Y, Z]$.
(2) Show that $\mathbb{C}[X, Y, Z]/I_2 \simeq \mathbb{C}[Z]$.
(3) Show that $\mathbb{C}[X, Y, Z]/I_3 \simeq \mathbb{C}$.
We have inclusions

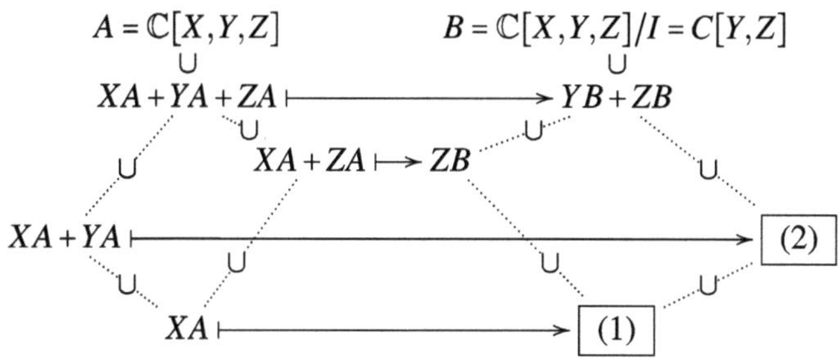

Fig. 10.3 Γ and Δ

$$\{0\} \subset I_1 \subset I_2 \subset I_3 \subset \mathbb{C}[X, Y, Z].$$

[6] Consider the ring $\mathbb{Z}$ and its ideal $I = 8\mathbb{Z}$. Draw a figure like the one in Example 10.30.

[7] Consider the ring $A = \mathbb{C}[X, Y, Z]$ and its ideal $I = X\,\mathbb{C}[X, Y, Z]$. Fill in the squares in Fig. 10.3.

[8] Let us define an $\mathbb{Z}$-module isomorphism

$$\varphi : \mathbb{Z}/((2^3 \times 3^2 \times 5)\mathbb{Z}) \to \mathbb{Z}/(2^3\mathbb{Z}) \oplus \mathbb{Z}/(3^2\mathbb{Z}) \oplus \mathbb{Z}/(5\mathbb{Z})$$

by

$$\varphi : x + (2^3 \times 3^2 \times 5)\mathbb{Z} \mapsto (x + 2^3\mathbb{Z}, x + 3^2\mathbb{Z}, x + 5\mathbb{Z}).$$

Find a concrete expression for

$$\varphi^{-1} : \mathbb{Z}/(2^3\mathbb{Z}) \oplus \mathbb{Z}/(3^2\mathbb{Z}) \oplus \mathbb{Z}/(5\mathbb{Z}) \to \mathbb{Z}/((2^3 \times 3^2 \times 5)\mathbb{Z}).$$

Chapter 11
Finite Fields and Polynomial Rings

We study polynomial rings over finite fields, their ideals, and factor rings.

11.1 Finite Rings and Finite Fields

Let us begin with the ring $\mathbb{Z}$ and its ideals.

Theorem 11.1 *All the ideals of the ring $\mathbb{Z}$ have the form $n\mathbb{Z}$ for some non-negative integer n.*

Proof For the trivial ideal $\{0\}$, take $n = 0$. Let I be any non-zero ideal. Let n be the minimum positive integer in I.

(1) For any $x \in I$, unique integers y and r satisfying

$$x = ny + r \quad (x, y, r \in \mathbb{Z}, 0 \le r \le n - 1)$$

are determined. Since $x, ny \in I$, we have

$$r = x - ny \in I,$$

which must be 0 from the minimality of n. Thus, we have

$$I \subset n\mathbb{Z}.$$

(2) Since $n \in I$ and I is closed under $\mathbb{Z}$-scalar multiplication, we have

$$n\mathbb{Z} \subset I.$$

S. Yukita, *First Lectures in Algebra*, Springer Asia Pacific Mathematics Series 10,
https://doi.org/10.1007/978-981-95-5457-7_11

From (1) and (2) we have

$$I = n\mathbb{Z}.$$

□

Remark 11.2 In the above proof, integer $x \in I$ can be either positive or negative. We give examples of both cases:

- When $x = 10$ and $n = 3$, we have the quotient $y = 3$ and remainder $r = 1$.
- When $x = -7$ and $n = 5$, we have the quotient $y = -2$ and remainder $r = 3$.

Theorem 11.3 *A natural number $n \geq 2$ is prime if and only if $n\mathbb{Z}$ is a prime ideal of $\mathbb{Z}$.*

Proof

(1) Assume n is not a prime number. We have $n = n_1 n_2$ for some natural numbers $n_1, n_2 \geq 2$. We have

$$\bar{n}_1 \neq \bar{0}, \quad \bar{n}_2 \neq \bar{0}, \quad \overline{n_1 n_2} = \bar{0}$$

in the factor ring $\mathbb{Z}/n\mathbb{Z}$, which means there exists a zero divisor in $\mathbb{Z}/n\mathbb{Z}$. By definition, $n\mathbb{Z}$ is not a prime ideal.

(2) Assume $n\mathbb{Z}$ is not a prime ideal of the ring $\mathbb{Z}$. The factor ring $\mathbb{Z}/n\mathbb{Z}$ has zero divisors $\bar{n}_1$ and $\bar{n}_2$, where we can take $n_1, n_2 \geq 2$. Since $\bar{n}_1 \bar{n}_2 = \overline{n_1 n_2} = \bar{0}$, there is a natural number m with $n_1 n_2 = nm$. If n is a prime number, then n divides n_1 or n_2. This means that $\bar{n}_1 = \bar{0}$ or $\bar{n}_2 = \bar{0}$. So, n must not be a prime number.

□

Theorem 11.4 *A natural number n is prime if and only if the $n\mathbb{Z}$ is a maximal ideal of the ring $\mathbb{Z}$.*

Proof

(1) Assume $n\mathbb{Z}$ is a maximal ideal. If the factor ring $\mathbb{Z}/n\mathbb{Z}$ has zero divisors $\bar{n}_1$ and $\bar{n}_2$ with $\bar{n}_1 \bar{n}_2 = \bar{0}$ and $2 \leq n_1, n_2 < n$. This means $n\mathbb{Z} \subsetneqq n_i \subsetneqq \mathbb{Z}$, which contradicts the assumption that $n\mathbb{Z}$ is maximal. Thus, $\mathbb{Z}/n\mathbb{Z}$ is an integral domain. By definition $n\mathbb{Z}$ is a prime ideal of $\mathbb{Z}$.

(2) Assume n is prime. If $n\mathbb{Z}$ is not a maximal ideal, then we have an ideal I such that $n\mathbb{Z} \subsetneqq I \subsetneqq \mathbb{Z}$. Theorem 11.1 tells us that $I = m\mathbb{Z}$ for some natural number m. Since $n \in m\mathbb{Z}$, there is a natural number x such that $n = mx$. We have $x \neq 1$ since $n \neq m$. This contradicts the assumption that n is prime.

□

Remark 11.5 Theorems 11.3 and 11.4 tell us that the notions of a prime ideal and maximal ideal are the same for the ring $\mathbb{Z}$. Recall that this is not always the case as we have seen in exercises in Chap. 10.

So far we established that the factor ring $\mathbb{Z}/n\mathbb{Z}$ is a field if and only if n is prime. Let p be a prime number. We denote the field $\mathbb{Z}/n\mathbb{Z}$ by $\mathbb{F}_p$.

Example 11.6 The addition and multiplication tables of $\mathbb{F}_2$ are given as follows:

$+$	$\bar{0}$	$\bar{1}$
$\bar{0}$	$\bar{0}$	$\bar{1}$
$\bar{1}$	$\bar{1}$	$\bar{0}$

$\times$	$\bar{1}$
$\bar{1}$	$\bar{1}$

Example 11.7 The addition and multiplication tables of $\mathbb{F}_3$ are given as follows. From now on, we omit the bars if there can be no confusion.

$+$	0	1	2
0	0	1	2
1	1	2	0
2	2	0	1

$\times$	1	2
1	1	2
2	2	1

The set of units forms a group under multiplication.

Example 11.8 The set of units of the factor ring $\mathbb{Z}/12\mathbb{Z} = \mathbb{Z}_{12}$ is given by

$$\{1, 5, 7, 11\}.$$

The multiplication table is as follows:

$\times$	1	5	7	11	inverse	order
1	1	5	7	11	1	1
5	5	1	11	7	5	2
7	7	11	1	5	7	2
11	11	7	5	1	11	2

We immediately see that the set of units is a group isomorphic to $K_4 = C_2 \times C_2$. This group is expressed in many ways as

$$\begin{aligned}(\mathbb{Z}/12\mathbb{Z})^\times &= \{\bar{1}, \bar{5}, \bar{7}, \overline{11}\} \\ &\cong \{1, 5\} \times \{1, 7\} \cong \{1, 5\} \times \{1, 11\} \cong \{1, 7\} \times \{1, 11\}.\end{aligned}$$

The set of zero divisors of $\mathbb{Z}/12\mathbb{Z} = \mathbb{Z}_{12}$ is

$$\{\bar{2}, \bar{3}, \bar{4}, \bar{6}, \bar{8}, \bar{9}, \overline{10}\}.$$

Set $\mathbb{Z}/12\mathbb{Z} = A$. The non-trivial ideals of the ring A are enumerated as

$$\bar{2}A, \quad \bar{3}A, \quad \bar{4}A, \quad \bar{6}A.$$

Noticing that

$$\bar{2}A = \overline{10}A, \quad \bar{3}A = \bar{9}A, \quad \bar{4}A = \bar{8}A,$$

we see that the zero divisors and ideals of the factor group are closely related.

In finite rings, zero divisors and units are simply related.

Theorem 11.9 *A non-zero element of a finite ring is either a unit or zero divisor exclusively.*

Remark 11.10 In an infinite ring $\mathbb{C}[X]$, the element X is neither a unit nor zero-divisor.

Proof of Theorem 11.9 Let non-zero elements be enumerated as $x_1 = 1,\ x_2,\ \cdots,\ x_n$.

(1) Suppose $x \neq 0$ is not a zero divisor. None of $x \cdot 1,\ xx_2,\ \ldots,\ xx_n$ is 0 since x is not a zero divisor. No two of them are equal. In fact, if

$$xx_k = xx_\ell \quad (k \neq \ell)$$

for $k \neq \ell$, then

$$x(x_k - x_\ell) = 0,$$

therefore x is a zero divisor. This contradicts the assumption that x is not a zero divisor. Now, we have the following set equality:

$$\{1,\ x_2,\ \ldots,\ x_n\} = \{x,\ xx_2,\ \ldots,\ xx_n\}.$$

We see that $1 = xx_k$ for some k. Thus, x is a unit.

(2) Suppose x is a unit. The equation $xy = 0$ yields $y = x^{-1}xy = 0$. This means that x is not a zero divisor.

□

Remark 11.11 The argument in the proof above depends on the following fact. Let A be a finite set. If a map $A \to A$ is injective, then it is also surjective, and vice versa. If we drop the finiteness condition, we can easily present a counter example.

Let $A = \{1, 2, 3, 4, \ldots\}$ and $\varphi : A \to A$ be multiplication by 2. We can easily see that φ is injective but not surjective.

Remark 11.12 A well-known theorem by Wedderburn claims every finite division ring is commutative, therefore a field. We do not prove it here. See any advanced text in the references. Obvious examples of non-commutative finite rings are $M_n(\mathbb{F}_p)$ for $n \geq 2$ and a prime p. All of them are non-commutative and have zero-divisors.

Given a ring A, we denote the set of all units of A by $A^\times$. The set $A^\times$ is a group under multiplication.

Example 11.8 tempts us to conjecture the following.

Theorem 11.13 *Let $\bar{I}$ be an ideal of the factor ring $A = \mathbb{Z}/n\mathbb{Z}$ such that $\{\bar{0}\} \subsetneqq \bar{I} \subsetneqq A$. There is a zero divisor $\bar{a}$ in A such that $\bar{I} = \bar{a}A$. The non-zero elements of $\bar{I}$ are all zero divisors of the ring A.*

Proof Once an ideal contains a unit, it must be the whole A. So, the ideal $\bar{I}$ has no units. By Theorem 11.9, the ideal $\bar{I}$ consists of zero divisors and $\bar{0}$.

Lemma 10.26 guarantees that $\bar{I}$ can be expressed as $\bar{a}A$ for some natural number a which should be a divisor of n. □

11.2 Polynomial Rings Over Finite Fields

We study polynomial rings over $\mathbb{F}_p$ for $p = 2, 3$. Euclidean division and the factor theorem play central roles.

Example 11.14 We denote the polynomial ring over $\mathbb{F}_2$ by $\mathbb{F}_2[X]$. Let us enumerate polynomials of degrees up to 3:

$$\begin{array}{ll} \text{degree } 0 & 0, \quad 1; \\ \text{degree } 1 & X, \quad X+1; \\ \text{degree } 2 & X^2, \quad X^2+1, \quad X^2+X+1; \\ \text{degree } 3 & X^3, \quad X^3+1, \quad X^3+X, \quad X^3+X+1, \quad X^3+X^2, \\ & X^3+X^2+1, \quad X^3+X^2+X, \quad X^3+X^2+X+1. \end{array}$$

We can continue the enumeration indefinitely.

Let us pairwise multiply polynomials of degree 1. We have the following:

$\times$	X	$X+1$
X	X^2	X^2+X
$X+1$	X^2+X	X^2+1

Note that

$$(X+1)^2 = X^2 + (1+1)X + 1 = X^2 + 1.$$

Notice that the polynomial X^2+X+1 does not appear in this table, which means that this polynomial cannot be factorized or, in other words, irreducible in $\mathbb{F}_2[X]$. This fact may be double-checked by the factor theorem. Set $f(X) = X^2 + X + 1$. Then, we have $f(0) = f(1) = 1$, which means that the quadratic equation $f(X) = 0$ have no solutions in $\mathbb{F}_2$.

Example 11.15 Let us consider an ideal $I = (X^2+X+1)\,\mathbb{F}_2[X]$ of the polynomial ring $\mathbb{F}_2[X]$. The factor ring is expressed as

$$\begin{aligned}\mathbb{F}_2[X]/I &= \{aX + b + I \mid a, b \in \mathbb{F}_2\}\\ &= \{\bar{0}, \bar{1}, \bar{X}, \overline{X+1}\}.\end{aligned}$$

We write $aX + b$ instead of $\overline{aX + b}$ if there is no risk of confusion. The addition table is given below.

$+$	0	1	X	$X+1$
0	0	1	X	$X+1$
1	1	0	$X+1$	X
X	X	$X+1$	0	1
$X+1$	$X+1$	X	1	0

The multiplication table is given below.

$\times$	1	X	$X+1$	*inverse*	*order*
1	1	X	$X+1$	1	1
X	X	$X+1$	1	$X+1$	3
$X+1$	$X+1$	1	X	X	3

We see the following:

(1) The factor ring $\mathbb{F}_2[X]/((X^2 + X + 1)\,\mathbb{F}_2[X])$ is a field.
(2) The group of units $\left(\mathbb{F}_2[X]/\ ((X^2 + X + 1)\,\mathbb{F}_2[X])\right)^\times$ is a cyclic group.

Example 11.16 Let us consider an ideal $I = (X^3 + 1)\,\mathbb{F}_2[X]$ of the polynomial ring $\mathbb{F}_2[X]$. The factor ring is expressed as

$$\mathbb{F}_2[X]/I = \{\bar{0}, \bar{1}, \bar{X}, \overline{X+1}, \overline{X^2}, \overline{X^2+1}, \overline{X^2+X}, \overline{X^2+X+1}\}.$$

Set $f(X) = X^3 + 1$. Since $f(1) = 0$, the factor theorem says $X - 1$ namely $X + 1$ divides $f(X)$. The quotient is obtained as follows:

$$\begin{array}{rl}
 & X^2 + X + 1 \\ \cline{2-2}
X+1\,\big) & X^3 \qquad\qquad + 1 \\
 & X^3 + X^2 \\ \cline{2-2}
 & \quad X^2 \\
 & \quad X^2 + X \\ \cline{2-2}
 & \qquad X + 1 \\
 & \qquad X + 1 \\ \cline{2-2}
 & \qquad\quad 0
\end{array}$$

Thus, we have

$$f(X) = (X+1)(X^2+X+1).$$

The factor X^2+X+1 cannot be further factorized over $\mathbb{F}_2$. Zero divisors of $\mathbb{F}_2[X]/I$ are

$$\overline{X+1}, \quad \overline{X^2+X+1} \quad \text{and}$$

their multiples. We enumerate all the multiples of these. Multiples of $\overline{X+1}$ are:

$$\bar{X}\cdot\overline{(X+1)} = \overline{X^2+X},$$
$$\overline{X+1}^2 = \overline{X^2+1}.$$

Multiples of $\overline{X^2+X+1}$ are:

$$\bar{X}\cdot\overline{X^2+X+1} = \overline{X^2+X+1},$$
$$\overline{X+1}\cdot\overline{X^2+X+1} = \bar{0},$$
$$\overline{X^2+X+1}^2 = \overline{X^2+X+1}.$$

To sum up, zero divisors of the factor ring $\mathbb{F}_2[X]/I$ are enumerated as

$$\overline{X+1}, \quad \overline{X^2+X+1}, \quad \overline{X^2+X}, \quad \overline{X^2+1}.$$

Theorem 11.9 says the group of units is

$$(\mathbb{F}_2[X]/I)^{\times} = \{\bar{1}, \bar{X}, \overline{X^2}\} \cong C_3,$$

obtained by excluding $\bar{0}$ and zero divisors from the ring $\mathbb{F}_2[X]/I$.

Example 11.17 Let us consider an ideal $I = (X^2+1)\,\mathbb{F}_3[X]$ of the polynomial ring $\mathbb{F}_3[X]$. The factor ring is expressed as

$$\mathbb{F}_3[X]/I = \{\bar{0}, \bar{1}, \bar{2}, \bar{X}, \overline{X+1}, \overline{X+2}, \overline{2X}, \overline{2X+1}, \overline{2X+2}\}.$$

Set $f(X) = X^2+1$. We have

$$f(0) = 1, \quad f(1) = 2, \quad \text{and} \quad f(2) = 2.$$

We see that X^2+1 cannot be factorized over $\mathbb{F}_3$. This means the factor ring has no zero divisors. By Theorem 11.9, we conclude $\mathbb{F}_3[X]/I$ is a field. The group of units is a cyclic group. The multiplication table is given as follows:

$\times$	1	2	X	$X+1$	$X+2$	$2X$	$2X+1$	$2X+2$	*order*
1	1	2	X	$X+1$	$X+2$	$2X$	$2X+1$	$2X+2$	1
2	2	1	$2X$	$2X+2$	$2X+1$	X	$X+2$	$X+1$	2
X	X	$2X$	2	$X+2$	$2X+2$	1	$X+1$	$2X+1$	4
$X+1$	$X+1$	$2X+2$	$X+2$	$2X$	1	$2X+1$	2	X	8
$X+2$	$X+2$	$2X+1$	$2X+2$	1	X	$X+1$	$2X$	2	8
$2X$	$2X$	X	1	$2X+1$	$X+1$	2	$2X+2$	$X+2$	4
$2X+1$	$2X+1$	$X+2$	$X+1$	2	$2X$	$2X+2$	X	1	8
$2X+2$	$2X+2$	$X+1$	$2X+1$	X	2	$X+2$	1	$2X$	8

11.3 Groups of Units of Finite Fields

We deal with some advanced topics. So, the reader may skip this section on first reading.

Various examples in the previous section tempt us to conjecture the following.

Theorem 11.18 *Let F be a finite field. The group of units $F^\times = F\backslash\{0\}$ is a cyclic group.*

This theorem is a direct consequence of the following more general theorem.

Theorem 11.19 *Let F be a field. If G is a finite subgroup of $F^\times$, then G is cyclic.*

We need some preparations.

Proposition 11.20 *Let G be a group and $a \in G$ be an element of finite order n. If d is a divisor of n, then the order of a^d is n/d.*

Proof Set $b = a^d$. If $b^m = 1$, then we have

$$b^m = (a^d)^m = a^{md} = 1,$$

which means $md \equiv 0 \mod n$. This is equivalent to

$$m \equiv 0 \mod n/d.$$

Now, we know the order of $b = a^d$ is n/d. □

Proposition 11.21 *Let G be a group and $a, b \in G$ be commutative elements of finite order m and n, respectively. If m and n are mutually coprime, then the order of ab is mn.*

Remark 11.22 The group G may or may not be abelian. The commutativity of elements a and b is essential. Recall that $\sigma = (1\ 2\ 3)$ and $\tau = (1\ 2)$ in group S_3 are of order 3 and 2, respectively. Their product $\sigma\tau$ has order 2.

Proof of Proposition 11.21 Let k be the order of ab. In general, we have no clue even to the finiteness of k. However, if we use the assumption $ab = ba$, then we have

$$(ab)^{mn} = a^{mn}b^{mn} = (a^m)^n(b^n)^m = e^n e^m = e,$$

which shows $k|mn$, namely k divides mn.

Since $(ab)^k = e$, we have

$$e = (ab)^{mk} = a^{mk}b^{mk} = (a^m)^k b^{mk} = eb^{mk} = b^{mk},$$

which shows $n|mk$. Since m and n are mutually coprime, we have $n|k$. Likewise, we have $m|k$. Since m and n are coprime, we have $mn|k$. Combining $k|mn$ and $mn|k$, we have $k = mn$. □

Proposition 11.23 *Let G be an abelian group. Let $a, b \in G$ have finite orders k and ℓ, respectively. Let m be the least common multiple of k and ℓ. There exist natural numbers s and t such that $a^s b^t$ has order m.*

Remark 11.24 The point is that we can construct an element of order m as a product of the powers of a and b.

We cannot hope $s = t = 1$, namely ab has order m, in general. Let $G = C_{30}$ be a cyclic group with generator g. Take $a = g^3$ and $b = g^5$ whose orders are 10 and 6, respectively. We know the least common multiple is 30. However, the order of $ab = g^8$ is 15.

Proof of Proposition 11.23 Let P be the union of the sets of prime factors of k and ℓ. The prime factorization of k and ℓ may be expressed as

$$k = \prod_{p \in P} p^{k_p}, \quad \ell = \prod_{p \in P} p^{\ell_p},$$

where $k_p \geq 0$ and $\ell_p \geq 0$. The least common multiple of k and ℓ is given by

$$m = \prod_{p \in P} p^{\max(k_p, \ell_p)}.$$

We adopt the following strategy:

(1) For each $p \in P$, we seek an element c_p of order $p^{\max(k_p,\ell_p)}$, where c_p is the product of some powers of a and b.
(2) For any different primes $p, q \in P$, the orders of c_p and c_q, namely $p^{\max(k_p,\ell_p)}$ and $q^{\max(k_q,\ell_q)}$ are mutually coprime. Using Proposition 11.21 repeatedly, we see that the order $\prod_{p \in P} c_p$ is $m = \prod_{p \in P} p^{\max(k_p,\ell_p)}$.
(3) The natural number $\prod_{p \in P} c_p$ must be the product of some powers of a and b.

We define c_p by

$$c_p = \begin{cases} a^{k/p^{k_p}} = a^{k/p^{\max(k_p,\ell_p)}} & \text{if } k_p \geq \ell_p \\ b^{\ell/p^{\ell_p}} = b^{\ell/p^{\max(k_p,\ell_p)}} & \text{if } k_p < \ell_p \end{cases}$$

for any $p \in P$. The exponents of a and b in the above expression are all non-negative integers since $p^{k_p}|k$ and $p^{\ell_p}|\ell$. By Proposition 11.20, we know that the order of c_p is $p^{\max(k_p,\ell_p)}$ in either case. Thus, we accomplished step (1). Steps (2) and (3) follow immediately. □

The proof above provides us with an algorithm. The following nearly concrete example may help to get an idea.

Example 11.25 Let G be an abelian group. Let a and b be elements of finite order

$$k = 2 \times 3^2 \times 5 \quad \text{and} \quad \ell = 2^2 \times 3 \times 5^2,$$

respectively. The least common multiple of k and ℓ is $m = 2^2 \times 3^2 \times 5^2$.

(1) Find c_2.
Since $k_2 = 1 < \ell_2 = 2$, we set

$$c_2 = b^{\ell/2^{\ell_2}} = b^{3\times 5^2},$$

whose order is 2^2.

(2) Find c_3.
Since $k_3 = 2 > \ell_3 = 1$, we set

$$c_3 = a^{k/3^{k_3}} = a^{2\times 5},$$

whose order is 3^2.

(3) Find c_5.
Since $k_5 = 1 < \ell_5 = 2$, we set

$$c_5 = b^{\ell/5^{\ell_5}} = b^{2^2\times 3},$$

whose order is 5^2.

Multiplying all these, we have

$$c_2 \times c_3 \times c_5 = a^{2\times 5} b^{3\times 5^2 + 2^2\times 3},$$

whose order is $m = 2^2 \times 3^2 \times 5^2$.

We add another example which is really concrete.

Example 11.26 We revisit Remark 11.24 and apply the algorithm presented in the proof of Proposition 11.23 to the example. Let $G = C_{30}$ be a cyclic group with a generator g. The elements $a = g^3$ and $b = g^5$ are of order 2×5 and 2×3, respectively. The least common multiple of their orders is $2 \times 3 \times 5 = 30$. The algorithm gives

$$c_2 = a^{3\times 5} \quad \text{or} \quad b^{3\times 5},$$
$$c_3 = a^{2\times 5},$$
$$c_5 = b^{2\times 3}..$$

If we choose the former of the "or," then, we have

$$c_2 \times c_3 \times c_5 = a^{25} b^6.$$

If we choose the latter, we have

$$c_2 \times c_3 \times c_5 = a^{10} b^{21}.$$

Both are equal to g^{23} whose order is 30.

Lemma 11.27 *Let F be a field. Let $a \in F^\times$ have finite order k. Then, we have*

$$\{x \in F^\times \mid x^k = 1\} = \{1, a, a^2, \cdots, a^{k-1}\}.$$

This means that the powers of a exhaust all elements of $F^\times$ with orders that are divisors of k.

Proof Any element of order that divides k satisfies $X^k = 1$. Take any element a^s from the RHS. Since $(a^s)^k = a^{sk} = (a^k)^s = 1$, the order of a^s is a divisor of k. Therefore, the RHS consists of elements of orders that are divisors of k. Thus, we have

$$\{x \in F^\times \mid x^k = 1\} \supset \{1, a, a^2, \cdots, a^{k-1}\}.$$

The divisor theorem says that the number of elements of the LHS is less than or equal to k. Therefore, we have the following equality:

$$\{x \in F^\times \mid x^k = 1\} = \{1, a, a^2, \cdots, a^{k-1}\}.$$

□

Proof of Theorem 11.19 Let n be the size of G. Let $a \in G$ be an element that has the largest order k. We prove that $k = n$.

Assume $k < n$. There must be $b \in G$ such that

$$b \notin \{1, a, a^2, \ldots, a^{k-1}\}.$$

Let ℓ be the order of b. Lemma 11.27 says that the order ℓ does not divide k. Let m be the least multiple k and ℓ. We see that $m > k$. Proposition 11.23 says that there is an element of order m, which contradicts the assumption that k is the largest order. □

Remark 11.28 Let F be a finite field. The group $F^\times$ is finite. So, we can apply Theorem 11.19 to $F^\times$. Thus, the group of units of a finite field is cyclic.

Finite fields of the same size are isomorphic. See Mullen and Mummert's book in the References.

Exercises

[1] Write down the multiplication table of each finite field of the following. Verify the group of units of each field is cyclic.

(1) $\mathbb{F}_5$.
(2) $\mathbb{F}_7$.
(3) $\mathbb{F}_{11}$.

[2] Write down the multiplication table of each finite ring of the following. Show that the groups of units are not cyclic.

(1) $(\mathbb{Z}/15\mathbb{Z})^\times$.
(2) $(\mathbb{Z}/8\mathbb{Z})^\times$.

[3] Consider the following ideals I_1 and I_2 of the polynomial ring $\mathbb{F}_2[X]$:

$$I_1 = (X^3 + X + 1)\mathbb{F}_2[X],$$
$$I_2 = (X^3 + X^2 + 1)\mathbb{F}_2[X].$$

(1) Write down the multiplication tables of $\mathbb{F}_2[X]/I_1$ and $\mathbb{F}_2[X]/I_2$. Verify that both factor rings are fields.
(2) Find a solution of $X^3 + X + 1 = 0$ in $\mathbb{F}_2[X]/I_2$. We express the solution by $a \in \mathbb{F}_2[X]/I_2$. Define a map $\varphi : \mathbb{F}_2[X]/I_1 \to \mathbb{F}_2[X]/I_2$ by

$$\varphi : \bar{X} \mapsto a.$$

Verify the map φ is an isomorphism of fields.

[4] Consider the following ideal I of the polynomial ring $\mathbb{F}_3[X]$,

$$I = (X^2 + X + 1)\mathbb{F}_3[X].$$

Find all zero divisors of the factor ring $\mathbb{F}_3[X]/I$. Write down the multiplication table of $(\mathbb{F}_3[X]/I)^\times$.

Chapter 12
Codes and Finite Fields

We will see an application of the finite fields, a gentle introduction to coding theory, which lays the firm foundation for digital communication technologies.

12.1 Geometry Over $\mathbb{F}_2$

If we replace $\mathbb{R}$ or $\mathbb{C}$ in the high-school geometry by finite fields, then what kind of geometry can we enjoy? The answers may be given by *algebraic geometry* or *arithmetic geometry*. In the following, we concentrate on lines and planes over $\mathbb{F}_2$.

The three-dimensional vector space over $\mathbb{F}_2$

$$\mathbb{F}_2{}^3 = \{(x, y, z) \mid x, y, z \in \mathbb{F}_2\}$$

consists of $2^3 = 8$ points. Planes in this space can be enumerated as follows:

$$\begin{aligned} x + y + z &= 0 \text{ or } 1, \\ x + y \phantom{{}+z} &= 0 \text{ or } 1, \\ x \phantom{{}+y} + z &= 0 \text{ or } 1, \\ x \phantom{{}+y+z} &= 0 \text{ or } 1, \\ y + z &= 0 \text{ or } 1, \\ y \phantom{{}+z} &= 0 \text{ or } 1, \\ z &= 0 \text{ or } 1. \end{aligned}$$

The total number of planes is 14. Every plane contains four points. For example, the plane $x + y + z = 1$ is viewed as

$$z = x + y + 1,$$

S. Yukita, *First Lectures in Algebra*, Springer Asia Pacific Mathematics Series 10,
https://doi.org/10.1007/978-981-95-5457-7_12

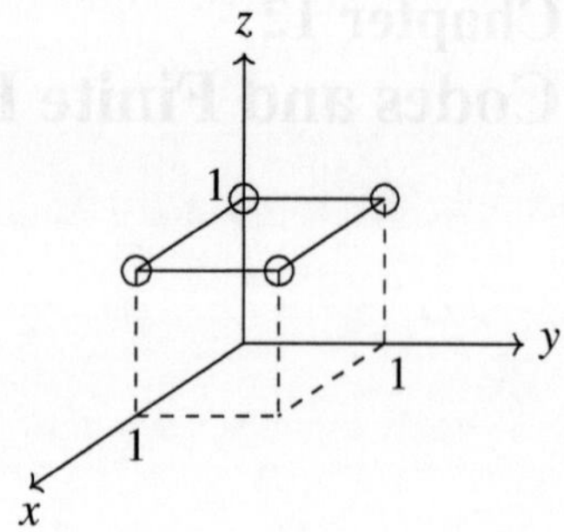

Fig. 12.1 Plane $z = 1$

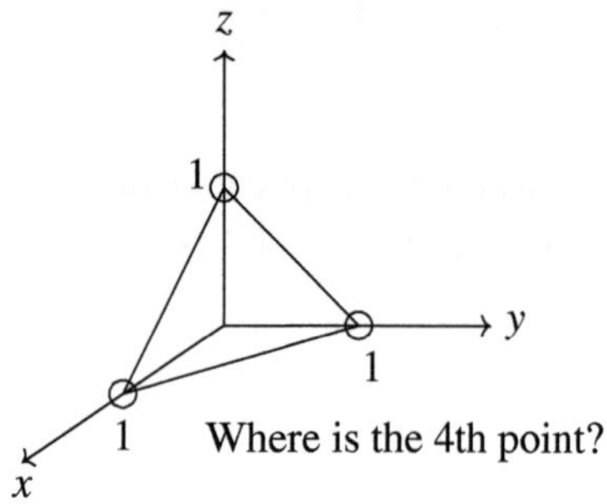

Fig. 12.2 Plane $x + y + z = 1$

which indicates that we can choose x and y independently on the plane. Thus, we have a total of $2 \times 2 = 4$ points on the plane.

Let us stimulate our intuition. There are four points $(0, 0, 1)$, $(0, 1, 1)$, $(1, 0, 1)$, and $(1, 1, 1)$ on the plane $z = 1$. See Fig. 12.1.

Consider the plane $x + y + z = 1$. We immediately get a picture as in Fig. 12.2, where three points $(1, 0, 0)$, $(0, 1, 0)$, and $(0, 0, 1)$ lie on the plane. Where has the fourth point gone? Calculation gives $(1, 1, 1)$. Analogy with the geometry of $\mathbb{R}^3$ may or may not assist the study of geometry over finite fields.

Next, we consider lines in the space ${\mathbb{F}_2}^3$. Using a parameter t, a line can be expressed as follows:

$$\begin{cases} x = pt + a \\ y = qt + b \\ z = rt + c \end{cases}.$$

The range of t is $\{\bar{0}, \bar{1}\} = \mathbb{F}_2$. Therefore, every line consists of two points. A line is determined by selecting two points. Thus, there are in total ${}_8C_2 = 28$ lines in the space ${\mathbb{F}_2}^3$.

Quiz 12.1 Describe a line that is the intersection of two planes $x + y + z = 0$ and $z = 0$.

12.2 Linear Codes and Ideals of Polynomial Rings

Reading and writing of digital data such as text, audio, and video is faced always with noise and errors arising from device interfaces or communication channels. In this chapter, we will see one of the most elementary techniques to fight those setbacks. The geometry over $\mathbb{F}_2$ just fits for this purpose.

In any data communication, there are disturbances. The coding techniques enable us to achieve two essential objectives:

(1) To detect errors caused by disturbances.
(2) To automatically correct those errors once detected.

Even if we abandon goal (2), the detection of errors, namely goal (1), alone may help us. When some errors are detected, the receiver requests retransmission from the sender. So, we can avoid miscommunication, a great achievement. Here, the receiver and sender are not restricted to humans, They may be devices. If goal (2) is also achieved, we are greatly relieved in time-critical situations.

Data communication lives in a world of bits. Every piece of information is considered as a point in ${\mathbb{F}_2}^n$. We introduce the distance function d for points in ${\mathbb{F}_2}^n$. For any two points

$$x = (x_1, x_2, \ldots, x_n), \quad y = (y_1, y_2, \ldots, y_n) \in {\mathbb{F}_2}^n,$$

we define $d(x, y)$ as the number of $i \in \{1, 2, \ldots, n\}$ such that $x_i \neq y_i$. This distance function is called the *Hamming distance*. For example, for $u = (1, 1, 1, 0)$, $v = (1, 0, 0, 0) \in {\mathbb{F}_2}^4$, we have $d(u, v) = 2$.

Quiz 12.2 Verify that the function d above satisfies the following axioms:

(1) $d(x, y) \geq 0$, where equality holds only when $x = y$.
(2) $d(x, y) + d(y, z) \geq d(x, z)$.

Detection and automatic correction of errors are accomplished by adding some redundancy to the data to be transmitted. We begin by an example of a brute-force approach that everyone may think of and must have put into practice in everyday life.

Example 12.1 (Method of Reconfirmation) Assume the sender wants to send 4-bit information. The receiver expects 4-bit data, one of possibly $2^4 = 16$ vectors in ${\mathbb{F}_2}^4$. The peers are afraid of errors. So, they decided to send and receive data with three times repetition. In other words, the sender embeds the piece of information of 4-bit into ${\mathbb{F}_2}^{12}$. We define an injection

$$\tau : {\mathbb{F}_2}^4 \to {\mathbb{F}_2}^{12}.$$

The most straightforward implementation is given by

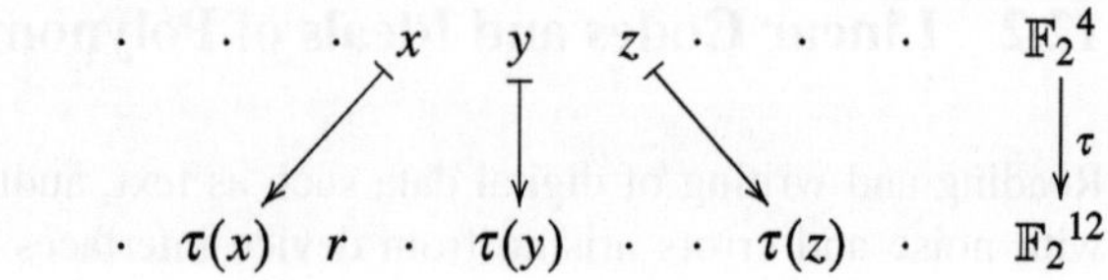

Fig. 12.3 At most one bit error

$$\tau : (x_1, x_2, x_3, x_4) \mapsto (x_1, x_2, x_3, x_4, x_1, x_2, x_3, x_4, x_1, x_2, x_3, x_4).$$

For any $x, y \in \mathbb{F}_2{}^4$ with $x \neq y$, we have

$$d(\tau(x), \tau(y)) \geq 3.$$

The map τ is called a *coding function*. Suppose that the sender wants to send a piece of information $x \in \mathbb{F}_2{}^4$. The sender transmits $\tau(x) \in \mathbb{F}_2{}^{12}$ into the communication channel. If there is no disturbance, the receiver gets $\tau(x) \in \mathbb{F}_2{}^{12}$. The subspace $\tau(\mathbb{F}_2{}^4) \subset \mathbb{F}_2{}^{12}$ has an important property that all the points are mutually separated by a distance of 3 or more. Assume that the quality of the channel guarantees the number of bit errors is at most one. Figure 12.3 shows the situation where the receiver got $r \in \mathbb{F}_2{}^{12}$ from the channel and seek the nearest point in the subspace $\tau(\mathbb{F}_2{}^4)$, in this case $\tau(x)$. The receiver decides the correct message was $x \in \mathbb{F}_2{}^4$. If the channel admits errors at most two bits, we need $d(\tau(x), \tau(y)) \geq 5$. This time we have to construct $\tau : \mathbb{F}_2{}^4 \to \mathbb{F}_2{}^{20}$.

So much for a brute-force approach. We seek a more efficient way using a *linear code*. The method of reconfirmation in Example 12.1 required 12 bits for communicating 4 bits. A new method requires 7 bits for communicating 4 bits, detecting, and correcting of errors at most 1 bit.

We want to construct an injective coding function

$$\tau : \mathbb{F}_2{}^4 \to \mathbb{F}_2{}^7.$$

We first summarize how to use it to attain the goals.

(1) The sender prepares a 4-bit vector $x \in \mathbb{F}_2{}^4$ that represents the piece of information to communicate.
(2) The sender emits $\tau(x) \in \mathbb{F}_2{}^7$ into the channel. The vector $\tau(x)$ is called a *code word*.
(3) The receiver gets $y \in \mathbb{F}_2{}^7$.
(4) If $y \in \operatorname{Im} \tau$, then the receiver finds the message as $x = \tau^{-1}(y)$. (Or we should write $x \in \tau^{-1}(\{y\})$ formally.) If $y \neq \operatorname{Im} \tau$, the receiver recognizes that the data y should not be accepted as it is.

Let us consider the factor ring

$$A = \mathbb{F}_2[X]/(X^7 - 1)\,\mathbb{F}_2[X],$$

which can be regarded as a 7-dimensional $\mathbb{F}_2$ vector space. An obvious basis is given by

$$\bar{1}, \overline{X}, \overline{X^2}, \ldots, \overline{X^6}.$$

We identify A and ${\mathbb{F}_2}^7$ as

$$\overline{a_0 + a_1 X + a_2 X^2 + \ldots + a_6 X^6} \quad \mapsto \quad (a_0, a_1, \ldots, a_6) \in {\mathbb{F}_2}^7.$$

Since $X^7 - 1$ is factorized as

$$X^7 - 1 = (X+1)(X^3 + X + 1)(X^3 + X^2 + 1),$$

we have an ideal

$$I = \overline{X^3 + X^2 + 1}\ A$$

of the ring A. The set I is a $\mathbb{F}_2$-linear subspace of ${\mathbb{F}_2}^7$. We want a basis of this subspace. Consider the following vectors:

$$\begin{aligned}
\overline{X^3 + X^2 + 1} \cdot \bar{1} &= \overline{X^3 + X^2 + 1} = \vec{f_0}, \\
\overline{X^3 + X^2 + 1} \cdot \overline{X} &= \overline{X^4 + X^3 + X} = \vec{f_1}, \\
\overline{X^3 + X^2 + 1} \cdot \overline{X^2} &= \overline{X^5 + X^4 + X^2} = \vec{f_2}, \\
\overline{X^3 + X^2 + 1} \cdot \overline{X^3} &= \overline{X^6 + X^5 + X^3} = \vec{f_3}.
\end{aligned}$$

We first show that $\vec{f_0}$, $\vec{f_1}$, $\vec{f_2}$, and $\vec{f_3}$ are linearly independent. Set

$$a_0 \vec{f_0} + a_1 \vec{f_1} + a_2 \vec{f_2} + a_3 \vec{f_3} = \vec{0}.$$

We immediately have

$$\begin{aligned}
&\overline{a_0 + a_1 X + (a_0 + a_2) X^2 + (a_0 + a_1 + a_3) X^3 + (a_1 + a_2) X^4 + (a_2 + a_3) X^5} \\
&\qquad \overline{+ a_3 X^6} = \bar{0}.
\end{aligned}$$

Since $1, X, X^2, \cdots, X^6$ are linearly independent, their coefficients must be zero. Solving the system of linear equations, we have

$$a_0 = a_1 = a_2 = a_3 = 0.$$

Further calculation gives the following. We leave the details to later exercises.

$$\overline{X^3+X^2+1}\cdot\overline{X^4}=\overline{X^7+X^6+X^4}$$
$$=\overline{X^6+X^4+1}$$
$$=\vec{f}_3+\vec{f}_2+\vec{f}_0, \tag{12.1}$$
$$\overline{X^3+X^2+1}\cdot\overline{X^5}=\overline{X^8+X^7+X^5}$$
$$=\overline{X^5+X+1}$$
$$=\vec{f}_2+\vec{f}_1+\vec{f}_0, \tag{12.2}$$
$$\overline{X^3+X^2+1}\cdot\overline{X^6}=\overline{X^9+X^8+X^6}$$
$$=\overline{X^6+X^2+X}$$
$$=\vec{f}_3+\vec{f}_2+\vec{f}_1. \tag{12.3}$$

All these results show that $\vec{f}_0$, $\vec{f}_1$, $\vec{f}_2$, and $\vec{f}_3$ form a basis of I as an $\mathbb{F}_2$ vector space.

Let us resume a postponed task of defining a coding function $\tau : {\mathbb{F}_2}^4 \to {\mathbb{F}_2}^7$. We set

$$\tau : (a_0, a_1, a_2, a_3) \mapsto \overline{X^3+X^2+1}\cdot\overline{a_0+a_1X+a_2X^2+a_3X^3},$$

or equivalently in the expanded form

$$\tau : (a_0, a_1, a_2, a_3) \mapsto$$
$$\overline{a_0+a_1X+(a_0+a_2)X^2+(a_0+a_1+a_3)X^3+(a_1+a_2)X^4+(a_2+a_3)X^5+a_3X^6}.$$

By the identification of A and ${\mathbb{F}_2}^7$, we can also express as

$$\tau : (a_0, a_1, a_2, a_3) \mapsto a_0\vec{f}_0+a_1\vec{f}_1+a_2\vec{f}_2+a_3\vec{f}_3.$$

We seek a matrix representation of the $\mathbb{F}_2$-linear map τ with respect to the basis

$$\begin{pmatrix}1\\0\\0\\0\end{pmatrix}, \begin{pmatrix}0\\1\\0\\0\end{pmatrix}, \begin{pmatrix}0\\0\\1\\0\end{pmatrix}, \begin{pmatrix}0\\0\\0\\1\end{pmatrix}$$

of ${\mathbb{F}_2}^4$ and the basis

$$\bar{1},\quad \overline{X},\quad \overline{X^2},\quad \ldots,\quad \overline{X^6}$$

of $A = {\mathbb{F}_2}^7$. The result is given by

$$\tau : \begin{pmatrix} a_0 \\ a_1 \\ a_2 \\ a_3 \end{pmatrix} \mapsto \begin{pmatrix} a_0 \\ a_1 \\ a_0 + a_2 \\ a_0 + a_1 + a_3 \\ a_1 + a_2 \\ a_2 + a_3 \\ a_3 \end{pmatrix} = \begin{pmatrix} 1\,0\,0\,0 \\ 0\,1\,0\,0 \\ 1\,0\,1\,0 \\ 1\,1\,0\,1 \\ 0\,1\,1\,0 \\ 0\,0\,1\,1 \\ 0\,0\,0\,1 \end{pmatrix} \begin{pmatrix} a_0 \\ a_1 \\ a_2 \\ a_3 \end{pmatrix}.$$

We consider next how to detect errors in a received vector in ${\mathbb{F}_2}^7$. Undisturbed data must be in $\operatorname{Im}\tau$. Let the received data be

$$\overline{f(X)} = \overline{b_0 + b_1 X + b_2 X^2 + \ldots + b_6 X^6} \in A = {\mathbb{F}_2}^7.$$

Proposition 12.2 *We have the following equivalence:*

$$\overline{f(X)} \in I \iff \overline{f(X)} \cdot \overline{X+1} \cdot \overline{X^3 + X + 1} = \overline{f(X)} \cdot \overline{X^4 + X^3 + X^2 + 1} = \bar{0}.$$

Proof

(1) We prove $\Longrightarrow$.
If $\overline{f(X)} \in I$, we have $\overline{f(X)} = \overline{X^3 + X^2 + 1} \cdot \overline{g(X)}$ for some $g(X) \in \mathbb{F}_2[X]$. Therefore,

$$\begin{aligned} &\overline{f(X)} \cdot \overline{X^4 + X^3 + X^2 + 1} \\ =&\overline{X^4 + X^3 + X^2 + 1} \cdot \overline{X^3 + X^2 + 1} \cdot \overline{g(X)} \\ =&\overline{X^7 - 1} \cdot \overline{g(X)} = \bar{0}. \end{aligned}$$

(2) We prove $\Longleftarrow$.
Assume

$$\overline{f(X)} \cdot \overline{X^4 + X^3 + X^2 + 1} = \bar{0}.$$

Divide $f(X)$ by $X^3 + X^2 + 1$ and get the quotient $h(X)$ and remainder $r(X)$ as follows:

$$\overline{f(X)} = \overline{X^3 + X^2 + 1} \cdot \overline{h(X)} + \overline{r(X)}.$$

Multiply $\overline{X^4 + X^3 + X^2 + 1}$ on both sides. By assumption we have

$$\bar{0} = \overline{X^4 + X^3 + X^2 + 1} \cdot \overline{r(X)}.$$

We must have $r(X) = 0$. Otherwise, the representative of the RHS

Fig. 12.4 $\operatorname{Im}\tau = \ker\sigma$

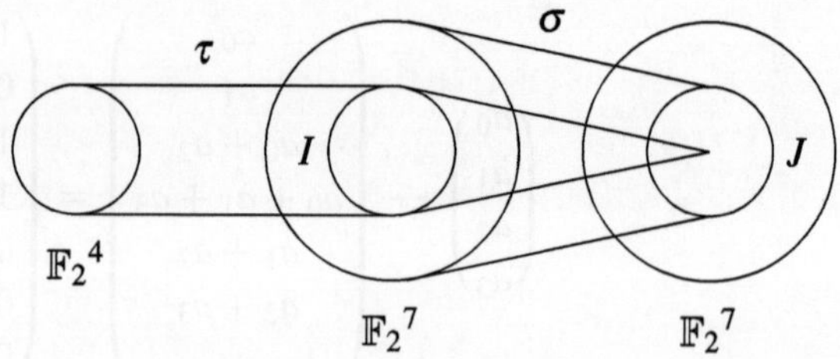

$$(X^4+X^3+X^2+1)\cdot r(X)$$

has degree 4, 5, or 6. This means the RHS is not zero, which is a contradiction. Thus, we have

$$\overline{f(X)} = \overline{X^3+X^2+1}\cdot\overline{h(X)},$$

which means $\overline{f(X)} \in I$.

□

We consider another ideal $J = \overline{X^4+X^3+X^2+1}\ A$. Note that J and I are complementary A-submodules of A. We define a map $\sigma : \mathbb{F}_2{}^7 = A \to J$ as

$$\sigma : \overline{f(X)} \mapsto \overline{f(X)}\cdot\overline{X^4+X^3+X^2+1}.$$

The receiver will use this map as follows. Let $\overline{f(X)}$ be the received vector. If $\sigma(\overline{f(X)}) = \bar{0}$, then we have $\overline{f(X)} \in \operatorname{Im}\tau$. Since τ is injective, we can extract the original message from the sender. We can summarize the argument so far as

$$\operatorname{Im}\tau = \ker\sigma.$$

The situation is illustrated in Fig. 12.4.

If $\overline{f(X)} \notin \ker\sigma$, namely $\overline{f(X)} \notin \operatorname{Im}\tau$, then what should we do? To answer this question, we have to study the ideal J a little bit. First, we seek a basis of J as a $\mathbb{F}_2$-vector space. Set

$$\overline{X^4+X^3+X^2+1}\cdot\bar{1} = \overline{X^4+X^3+X^2+1} = \vec{e_0},$$

$$\overline{X^4+X^3+X^2+1}\cdot\overline{X} = \overline{X^5+X^4+X^3+X} = \vec{e_1},$$

$$\overline{X^4+X^3+X^2+1}\cdot\overline{X^2} = \overline{X^6+X^5+X^4+X^2} = \vec{e_2}.$$

The vectors $\vec{e_0}, \vec{e_1},$ and $\vec{e_2}$ form a basis of J.

Quiz 12.3 Show that $\vec{e_0}, \vec{e_1},$ and $\vec{e_2}$ are $\mathbb{F}_2$-linearly independent.

Quiz 12.4 Verify the following equalities:

$$\overline{X^4 + X^3 + X^2 + 1} \cdot \overline{X^3} = \vec{e_1} + \vec{e_3},$$

$$\overline{X^4 + X^3 + X^2 + 1} \cdot \overline{X^4} = \vec{e_1} + \vec{e_2} + \vec{e_3},$$

$$\overline{X^4 + X^3 + X^2 + 1} \cdot \overline{X^5} = \vec{e_1} + \vec{e_2},$$

$$\overline{X^4 + X^3 + X^2 + 1} \cdot \overline{X^6} = \vec{e_2} + \vec{e_3}.$$

The matrix representation of σ with respect to the basis $\bar{1}, \overline{X}, \cdots, \overline{X^6}$ of $\mathbb{F}_2{}^7$ and the basis $\vec{e_0}, \vec{e_1}, \vec{e_2}$ of J is given by

$$\sigma : \begin{pmatrix} a_0 \\ \vdots \\ a_6 \end{pmatrix} \mapsto \begin{pmatrix} 1\,0\,0\,1\,1\,1\,0 \\ 0\,1\,0\,0\,1\,1\,1 \\ 0\,0\,1\,1\,1\,0\,1 \end{pmatrix} \begin{pmatrix} a_0 \\ \vdots \\ a_6 \end{pmatrix}.$$

The 3×7 matrix on the RHS is called a *parity check matrix*. The column vectors are pairwise linearly independent.

Proposition 12.3 *For any $x \neq y \in \operatorname{Im} \tau$, we have $d(x, y) \geq 3$.*

Proof Assume that $d(x, y) \leq 2$. The components of the vector $x - y$ are all 0's except for two, say the i, j-th components. Since $x, y \in \operatorname{Im} \tau$, we have $x - y \in \operatorname{Im} \tau$. We have $x - y \in \ker \sigma$ by the equality $\operatorname{Im} \tau = \ker \sigma$. Let P_i and P_j be the i, j-th column vectors of the parity check matrix. Since $\sigma(x - y) = 0$, we have

$$(x_i - y_i)P_i + (x_j - y_j)P_j = 0.$$

Linear independence of P_i and P_j gives $x_i = y_i$ and $x_j = y_j$. This contradicts the assumption $x \neq y$. □

All the points in $\operatorname{Im} \tau$ are mutually separated by a distance of 3 or more. This fact allows us to adopt the same strategy as the reconfirmation method, reducing the cost of bits from 12 to 7.

We consider next the method of retrieving the original 4-bit message from the received raw data, a vector in $\mathbb{F}_2{}^7$. Let $\overline{f(X)} \in A = \mathbb{F}_2{}^7$ be the received data. Let $\overline{h(X)}$ be the original message.

(1) The original message was first cast into $\mathbb{F}_2{}^7$ by τ as

$$\overline{h(X)} \cdot \overline{X^3 + X^2 + 1}.$$

(2) Let the received data be $\overline{f(X)}$.

   (i) If there is no disturbance, then the receiver gets

$$\overline{f(X)} = \overline{h(X)} \cdot \overline{X^3 + X^2 + 1}$$

as it is.

(ii) If the $(k+1)$-th bit was flipped by noise in the channel, then the received data should be

$$\overline{f(X)} = \overline{h(X)} \cdot \overline{X^3 + X^2 + 1} + \overline{X^k}.$$

(3) Multiply $\overline{f(X)}$ by $\overline{X^4 + X^3 + X^2 + 1}$.

(i) In the case of no disturbance, we have

$$\sigma(\overline{f(X)}) = \overline{h(X)} \cdot \overline{X^3 + X^2 + 1} \cdot \overline{X^4 + X^3 + X^2 + 1} = \bar{0}.$$

(ii) In the case of the $(k+1)$-th bit being flipped, we have

$$\sigma(\overline{f(X)}) = \overline{X^k} \cdot \overline{X^4 + X^3 + X^2 + 1} = \sigma(\overline{X^k}).$$

The term $\overline{X^k}$ is identified with the column vector with components 0 except for 1 with the $(k+1)$-th entry. The above expression can be rewritten in the matrix form as follows:

$$\begin{pmatrix} 1\,0\,0\,1\,1\,1\,0 \\ 0\,1\,0\,0\,1\,1\,1 \\ 0\,0\,1\,1\,1\,0\,1 \end{pmatrix} \begin{pmatrix} 0 \\ \vdots \\ 0 \\ 1 \\ 0 \\ \vdots \\ 0 \end{pmatrix} \begin{matrix} \\ \\ \\ \leftarrow \text{ the } (k+1)\text{-th entry,} \\ \\ \\ \\ \end{matrix}$$

which results in the $(k+1)$-th column vector of the parity check matrix.

(4) To extract the original message, apply the following correction to $\overline{f(X)}$ then divide by $X^3 + X^2 + 1$:

(i) In the case of no disturbance, no correction needed.

(ii) If $\sigma(\overline{f(X)}) = \sigma(\overline{X^k})$, replace $\overline{f(X)}$ by $\overline{f(X)} + \overline{X^k}$.

We walk through the process in a concrete example.

Example 12.4 Let the original message be $\begin{pmatrix} 1 \\ 1 \\ 0 \\ 0 \end{pmatrix}$. This data is encoded by τ as

$$\tau\begin{pmatrix}1\\1\\0\\0\end{pmatrix}=\begin{pmatrix}1\\1\\1\\0\\1\\0\\0\end{pmatrix}\in\mathbb{F}_2{}^7.$$

If there is no disturbance, this vector is received as it is. The reader should verify that this vector multiplied by the party check matrix from the left is the null vector. Assume that the noise flips the 4th bit. The received data should be

$$\text{The 4th bit is flipped}\rightarrow\begin{pmatrix}1\\1\\1\\1\\1\\0\\0\end{pmatrix}\in\mathbb{F}_2{}^7.$$

If we multiply the parity check matrix from the left, we have $\begin{pmatrix}1\\0\\1\end{pmatrix}$. This vector is exactly the 4th column of the parity check matrix. So, the receiver knows that the data has an error in the 4th bit. The corrected data is

$$\overline{1+X+X^2+X^4}\in\operatorname{Im}\tau\subset A.$$

To extract the original message $\tau^{-1}(\overline{1+X+X^2+X^4})$, we calculate the following division:

$$\begin{array}{r|l}
 & X+1 \\
\hline
X^3+X^2+1 & X^4 \qquad\qquad +X^2+X+1 \\
 & X^4+X^3 \qquad\qquad +X \\
\hline
 & \qquad X^3+X^2 \qquad\qquad +1 \\
 & \qquad X^3+X^2 \qquad\qquad +1 \\
\hline
 & \qquad\qquad\qquad\qquad\quad 0
\end{array}$$

The quotient $\overline{1+X}$ shows that the original message is $\begin{pmatrix}1\\1\\0\\0\end{pmatrix}$.

Usually, all these coding and decoding are implemented and embedded in hardware devices. Further studies on efficient coding are continuing. The reader may search the latest achievements with the key words “coding theory” on the Internet.

Exercises

[1] We study the geometry of $\mathbb{F}_3{}^3$.

(1) Count the number of all planes in $\mathbb{F}_3{}^3$.
(2) Count the number of all straight lines in $\mathbb{F}_3{}^3$.

[2] We study the geometry of $\mathbb{F}_2{}^4$.

(1) Count the number of all hyperplanes in $\mathbb{F}_2{}^4$. A hyperplane is a geometrical object that is defined by a single linear equation.
(2) Count the number of all straight lines in $\mathbb{F}_2{}^4$.

[3] We study irreducible polynomials of $\mathbb{F}_2[X]$.

(1) Enumerate polynomials of degree 1.
(2) Enumerate irreducible polynomials of degree 2.
(3) Enumerate irreducible polynomials of degree 3.
(4) Enumerate irreducible polynomials of degree 4.

[4] Veryfy Eqs. (12.1), (12.2), and (12.3). Namely, set

$$\vec{f_0} = \overline{X^3 + X^2 + 1},$$

$$\vec{f_1} = \overline{X^4 + X^3 + X},$$

$$\vec{f_2} = \overline{X^5 + X^4 + X^2},$$

$$\vec{f_3} = \overline{X^6 + X^5 + X^3}.$$

and show the following:

$$\overline{X^6 + X^4 + 1} = \vec{f_3} + \vec{f_2} + \vec{f_0},$$

$$\overline{X^5 + X + 1} = \vec{f_2} + \vec{f_1} + \vec{f_0},$$

$$\overline{X^6 + X^2 + X} = \vec{f_3} + \vec{f_2} + \vec{f_1}.$$

Appendix A
Solutions to Quizzes and Exercises

Chapter 1
Quiz 1.1 We fix x in $xa + y$ and vary y through $\mathbb{Z}$. Since xa is constant, $xa + y$ runs all elements of $\mathbb{Z}$.
Quiz 1.2 Let k be the least natural number such that $t_n = 0$ for all $n > k$. Let k' be the least natural number such that $s_n = 0$ for all $n > k'$. Let us retake k as the maximum of k and k'.

Chapter 2
Quiz 2.1 Let the transformation be

$$\begin{pmatrix} x \\ y \end{pmatrix} \mapsto \begin{pmatrix} x' \\ y' \end{pmatrix}.$$

Since the y-coordinate is preserved, we have $y' = y$. We can concentrate on the x-coordinate. We solve the linear equation

$$\frac{1}{2}(x + x') = b$$

with the unknown x'. The solution is

$$x' = 2b - x.$$

Quiz 2.2 We prove the two equations pointwise:

$$\begin{aligned} (R_b \circ X_a)\begin{pmatrix} x \\ y \end{pmatrix} &= R_b\left(X_a\begin{pmatrix} x \\ y \end{pmatrix}\right) \\ &= R_b\begin{pmatrix} x + a \\ y \end{pmatrix} \end{aligned}$$

S. Yukita, *First Lectures in Algebra*, Springer Asia Pacific Mathematics Series 10,
https://doi.org/10.1007/978-981-95-5457-7

$$= \begin{pmatrix} 2b - (x + a) \\ y \end{pmatrix}$$

$$= \begin{pmatrix} 2 \times \frac{2b-a}{2} - x \\ y \end{pmatrix}$$

$$= R_{\frac{2b-a}{2}} \begin{pmatrix} x \\ y \end{pmatrix},$$

$$\begin{aligned} (R_b \circ R_{b'}) \begin{pmatrix} x \\ y \end{pmatrix} &= R_b \left(R_{b'} \begin{pmatrix} x \\ y \end{pmatrix} \right) \\ &= R_b \begin{pmatrix} 2b' - x \\ y \end{pmatrix} \\ &= \begin{pmatrix} 2b - (2b' - x) \\ y \end{pmatrix} \\ &= \begin{pmatrix} 2b - 2b' + x \\ y \end{pmatrix} \\ &= X_{2b-2b'} \begin{pmatrix} x \\ y \end{pmatrix}. \end{aligned}$$

Quiz 2.3 We have the formulas of composition:

$$X_a \circ R_b = R_{\frac{a+2b}{2}}, \quad R_b \circ X_a = R_{\frac{2b-a}{2}}.$$

Set

$$R_{\frac{a+2b}{2}} = R_{\frac{2b-a}{2}}.$$

Both sides should be the same reflection. So, we have

$$\frac{a+2b}{2} = \frac{2b-a}{2},$$

which yields $a = 0$. Namely, we have

$$X_a \circ R_b = R_b \circ X_a \Leftrightarrow a = 0.$$

Quiz 2.4 We can easily verify the closedness of composition by writing down the table as follows:

	X_b	$X_b \circ S$
X_a	X_{a+b}	$X_{a+b} \circ S$
$X_a \circ S$	$X_{a+b} \circ S$	X_{a+b}

The identity transformation is X_0. The inverses are given by

$$(X_a)^{-1} = X_{-a}, \quad (X_a \circ S)^{-1} = X_{-a} \circ S.$$

Associativity is automatic since all are transformations.

Quiz 2.5 The composition formula is essentially the addition theorem of trigonometric functions:

$$\begin{aligned}
&\begin{pmatrix} \cos\alpha & -\sin\alpha \\ \sin\alpha & \cos\alpha \end{pmatrix} \begin{pmatrix} \cos\beta & -\sin\beta \\ \sin\beta & \cos\beta \end{pmatrix} \\
&= \begin{pmatrix} \cos\alpha\cos\beta - \sin\alpha\sin\beta & -(\sin\alpha\cos\beta + \sin\beta\cos\alpha) \\ \sin\alpha\cos\beta + \sin\beta\cos\alpha & \cos\alpha\cos\beta - \sin\alpha\sin\beta \end{pmatrix} \\
&= \begin{pmatrix} \cos(\alpha+\beta) & -\sin(\alpha+\beta) \\ \sin(\alpha+\beta) & \cos(\alpha+\beta) \end{pmatrix},
\end{aligned}$$

which shows the closedness of composition. The identity transformation is obtained by taking $\theta = 0$. The inverses are given by

$$\begin{pmatrix} \cos\theta & -\sin\theta \\ \sin\theta & \cos\theta \end{pmatrix}^{-1} = \begin{pmatrix} \cos(-\theta) & -\sin(-\theta) \\ \sin(-\theta) & \cos(-\theta) \end{pmatrix}.$$

Associativity is automatic since composition is matrix multiplication.

Quiz 2.6 Closedness is verified by

$$\begin{pmatrix} a & 0 \\ 0 & a^{-1} \end{pmatrix} \begin{pmatrix} b & 0 \\ 0 & b^{-1} \end{pmatrix} = \begin{pmatrix} ab & 0 \\ 0 & (ab)^{-1} \end{pmatrix}.$$

Take $a = 1$ for the identity transformation. Inverses are given by

$$\begin{pmatrix} a & 0 \\ 0 & a^{-1} \end{pmatrix}^{-1} = \begin{pmatrix} a^{-1} & 0 \\ 0 & a \end{pmatrix}.$$

Associativity is automatic since composition is matrix multiplication.

Quiz 2.7 We have seen that the positive part of the x-axis is one of the B-orbits. The B-orbit passing through $(-1, 0)$ is given by

$$\begin{pmatrix} a & 0 \\ 0 & a^{-1} \end{pmatrix} \begin{pmatrix} -1 \\ 0 \end{pmatrix} = \begin{pmatrix} -a \\ 0 \end{pmatrix},$$

which exactly covers the negative part of the x-axis when a runs all positive real numbers. Likewise, the positive part of the y-axis and the negative part of the y-axis are obtained as B-orbits passing through $(0, 1)$ and $(0, -1)$, respectively. The origin is a fixed point, which means the origin is itself a B-orbit.

Exercises

[1]

(1) $\{X_a, R_b \mid a \in 2\mathbb{Z}, b \in \mathbb{Z}\}$, where $2\mathbb{Z} = \{2n \mid n \in \mathbb{Z}\}$.

(2) $\{X_a, R_b, X_a \circ S_0, R_b \circ S_0 \mid a \in 2\mathbb{Z}, b \in \mathbb{Z}\}$.

Note that the set $\{X_a, R_b, S_0 \mid a \in 2\mathbb{Z}, b \in \mathbb{Z}\}$ is not closed. In Sect. 2.1 we got the following:

$$\begin{aligned} X_a \circ X_{a'} &= X_{a'} \circ X_a = X_{a+a'}, \\ R_{b'} \circ R_b &= X_{2b'-2b}, \\ X_a \circ R_b &= R_{\frac{a+2b}{2}}, \\ R_b \circ X_a &= R_{\frac{2b-a}{2}}. \end{aligned}$$

Since $a, a' \in 2\mathbb{Z}$ and $b, b' \in \mathbb{Z}$, we have

$$a + a' \in 2\mathbb{Z}, \quad 2b' - 2b \in 2\mathbb{Z}$$

for indices of X. We also have

$$\frac{a+2b}{2}, \quad \frac{2b-a}{2} \in \mathbb{Z}$$

for indices of R. Let us calculate compositions among X, R, and S_0:

$$\begin{aligned} X_a \circ S_0 \begin{pmatrix} x \\ y \end{pmatrix} &= X_a \begin{pmatrix} x \\ -y \end{pmatrix} = \begin{pmatrix} x+a \\ -y \end{pmatrix}, \\ S_0 \circ X_a \begin{pmatrix} x \\ y \end{pmatrix} &= S_0 \begin{pmatrix} x+a \\ y \end{pmatrix} = \begin{pmatrix} x+a \\ -y \end{pmatrix}, \\ R_b \circ S_0 \begin{pmatrix} x \\ y \end{pmatrix} &= R_b \begin{pmatrix} x \\ -y \end{pmatrix} = \begin{pmatrix} 2b-x \\ -y \end{pmatrix}, \\ S_0 \circ R_b \begin{pmatrix} x \\ y \end{pmatrix} &= S_0 \begin{pmatrix} 2b-x \\ y \end{pmatrix} = \begin{pmatrix} 2b-x \\ -y \end{pmatrix}. \end{aligned}$$

We see that S_0 commutes with both X_a and R_b.Let us examine the closedness of the set

$$\{X_a, R_b, S_0 \mid a \in 2\mathbb{Z}, b \in \mathbb{Z}\}.$$

The elements $X_a \circ S_0$ $(a \neq 0)$ and $R_b \circ S_0$ are excluded from the set. Incorporating these elements, we have the multiplication Table A.1, which shows that the set

$$\{X_a, R_b, X_a \circ S_0, R_b \circ S_0 \mid a \in 2\mathbb{Z}, b \in \mathbb{Z}\}$$

is closed under multiplication.

Table A.1 Multiplication table

Left \ Right	$X_{a'}$	$R_{b'}$	$X_{a'} \circ S_0$	$R_{b'} \circ S_0$
X_a	$X_{a+a'}$	$R_{\frac{a+2b'}{2}}$	$X_{a+a'} \circ S_0$	$R_{\frac{a+2b'}{2}} \circ S_0$
R_b	$R_{\frac{2b-a'}{2}}$	$X_{2b-2b'}$	$R_{\frac{2b-a'}{2}} \circ S_0$	$X_{2b-2b'} \circ S_0$
$X_a \circ S_0$	$X_{a+a'} \circ S_0$	$R_{\frac{a+2b'}{2}} \circ S_0$	$X_{a+a'}$	$R_{\frac{a+2b'}{2}}$
$R_b \circ S_0$	$R_{\frac{2b-a'}{2}} \circ S_0$	$X_{2b-2b'} \circ S_0$	$R_{\frac{2b-a'}{2}}$	$X_{2b-2b'}$

(3) $\{Y_a, S_a, Y_a \circ R_0, S_a \circ R_0 \mid a \in \mathbb{R}\}$.
Closedness is verified by Table A.2.

Table A.2 Multiplication table

Left \ Right	$Y_{a'}$	$S_{a'}$	$Y_{a'} \circ R_0$	$S_{a'} \circ R_0$
Y_a	$Y_{a+a'}$	$S_{\frac{a+2b'}{2}}$	$Y_{a+a'} \circ R_0$	$S_{\frac{a+2b'}{2}} \circ R_0$
S_a	$S_{\frac{2a-a'}{2}}$	$Y_{2a-2a'}$	$S_{\frac{2a-a'}{2}} \circ R_0$	$Y_{2a-2a'} \circ R_0$
$Y_a \circ R_0$	$Y_{a+a'} \circ R_0$	$S_{\frac{a+2a'}{2}} \circ R_0$	$Y_{a+a'}$	$S_{\frac{a+2a'}{2}}$
$S_a \circ R_0$	$S_{\frac{2a-a'}{2}} \circ R_0$	$Y_{2a-2a'}$	$S_{\frac{2a-a'}{2}}$	$Y_{2a-2a'}$

(4) $\{Y_a, Y_a \circ R_{\frac{3}{2}} \mid a \in \mathbb{Z}\}$.
(5) $\{I, R_0, S_0, R_0 \circ S_0\}$.

[2] Various solutions are possible.

(1) Place $\wedge$ signs at all points whose y-coordinates are even numbers on the line $x = -1$. Place, in addition, ◯ signs at all points whose y-coordinates are even numbers on the line $x = 1$. See Fig. A.1.
(2) Place □ signs at all points whose y-coordinates are even numbers on the line $x = -1$. Place, in addition, ◯ signs at all points whose y-coordinates are even numbers on the line $x = 1$. See Fig. A.2.

(3) Place $\wedge$ signs at all points whose y-coordinates are even numbers on the lines $x = \pm 1$. See Fig. A.3.

Fig. A.1 $\wedge$ and ○

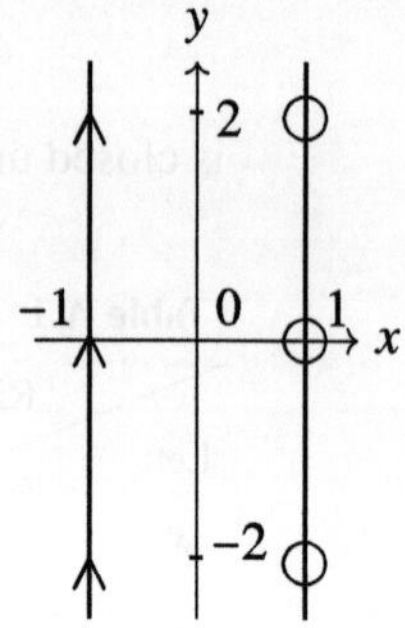

[3]

(1)

$$Z_{\frac{2\pi}{3}} \circ S_0 = \begin{pmatrix} \cos\frac{2\pi}{3} & \sin\frac{2\pi}{3} \\ \sin\frac{2\pi}{3} & -\cos\frac{2\pi}{3} \end{pmatrix} = S_0 \circ Z_{\frac{4\pi}{3}},$$

Fig. A.2 □ and ○

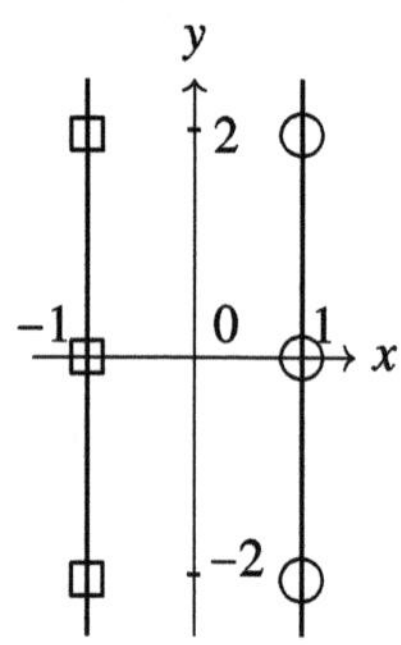

Fig. A.3 $\wedge$

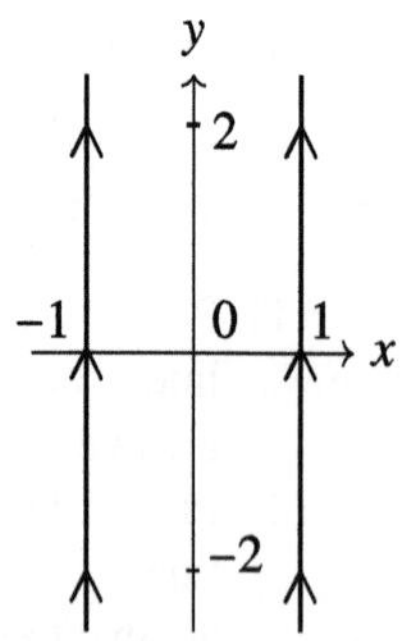

$$Z_{\frac{4\pi}{3}} \circ S_0 = \begin{pmatrix} \cos\frac{4\pi}{3} & \sin\frac{4\pi}{3} \\ \sin\frac{4\pi}{3} & -\cos\frac{4\pi}{3} \end{pmatrix} = S_0 \circ Z_{\frac{2\pi}{3}}.$$

(2)

$$\{I, Z_{\frac{2\pi}{3}}, Z_{\frac{4\pi}{3}}, S_0, Z_{\frac{2\pi}{3}} \circ S_0, Z_{\frac{4\pi}{3}} \circ S_0\}.$$

[4]

(1) $\{Z_{\frac{n\pi}{3}}, Z_{\frac{n\pi}{3}} \circ S_0 \,|\, n = 0, 1, \cdots, 5\}$. Since $Z_{\frac{n\pi}{3}} \circ S_0 = S_0 \circ Z_{-\frac{n\pi}{3}}$, we see that the set is closed under multiplication.

(2) $\{Z_{\frac{n\pi}{3}} \,|\, n = 0, 1, \cdots, 5\}$.

[5]

(1) See Figs. A.4 and A.5.

Fig. A.4 *A*-orbit

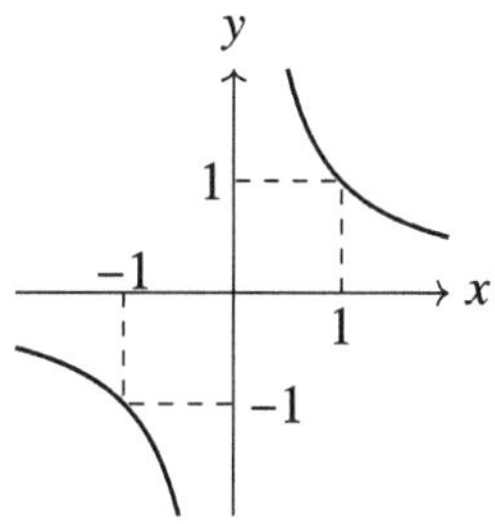

(2) See Figs. A.6 and A.7.

[6]

(1)

$$\left\{ \begin{pmatrix} a & 0 \\ 0 & b \end{pmatrix} \middle| a, b \in \mathbb{R}, \quad ab = \pm 1 \right\}.$$

Fig. A.5 *B*-orbit

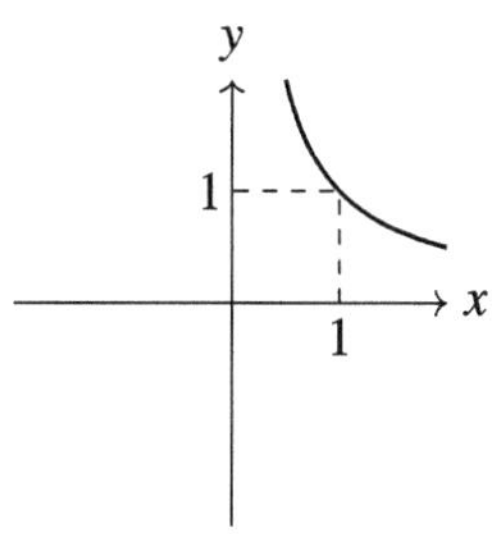

Fig. A.6 A-orbit

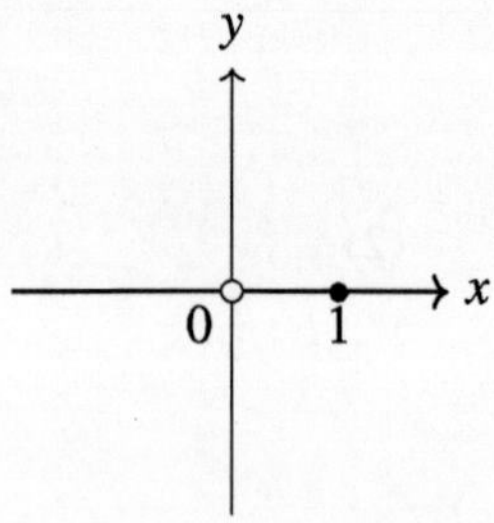

Fig. A.7 B-orbit

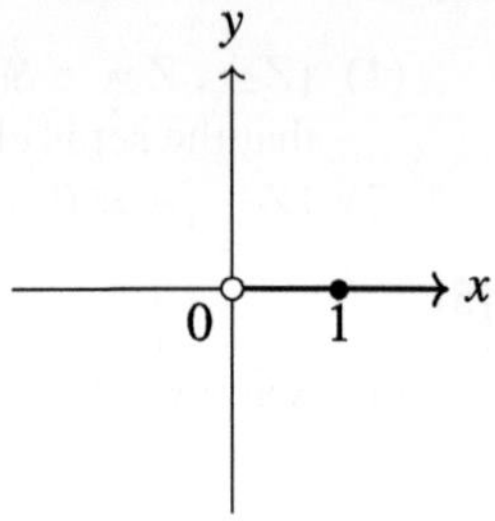

(2)

$$\left\{ \begin{pmatrix} a & 0 \\ 0 & b \end{pmatrix} \middle| a, b \in \mathbb{R}, \quad ab = 1 \right\}.$$

(3)

$$\left\{ \begin{pmatrix} a & 0 \\ 0 & b \end{pmatrix} \middle| a, b \in \mathbb{R}, \quad a > 0, \quad b \neq 0, \quad ab = \pm 1 \right\}.$$

[7]

(1) Three vertices of the regular triangle in Fig. A.8.
(2) Six vertices of the regular hexagon in Fig. A.9.
(3) Circle in Fig. A.10.

Fig. A.8 3 points

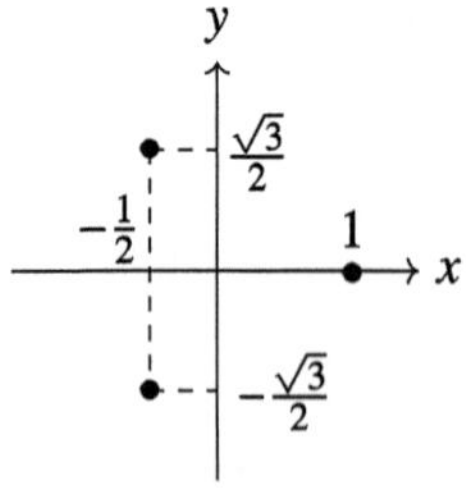

Fig. A.9 6 points

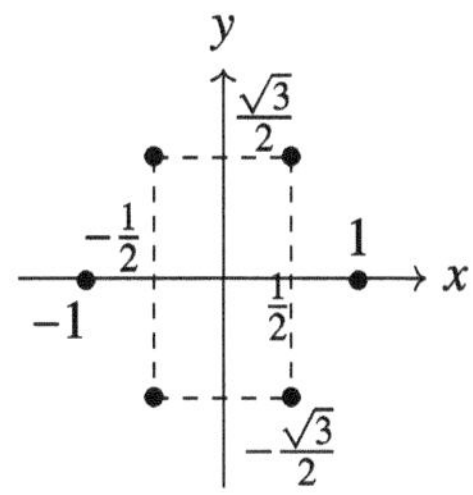

Fig. A.10 ∞ points

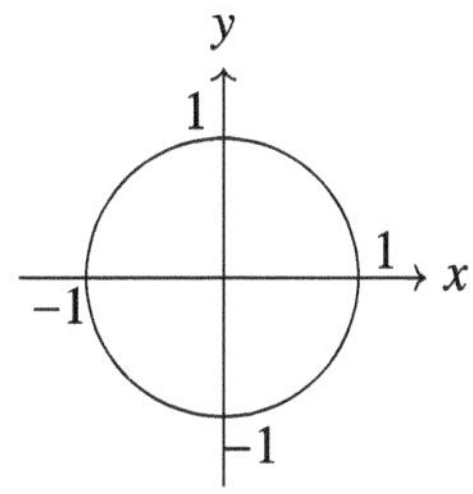

Chapter 3

Quiz 3.1 The identity element is $\begin{pmatrix} 1 & 2 & 3 \\ 1 & 2 & 3 \end{pmatrix}$. An example calculating an inverse is as follows. The inverse of the map $1 \mapsto 2,\ 2 \mapsto 3,\ 3 \mapsto 1$ is $2 \mapsto 1,\ 3 \mapsto 2,\ 1 \mapsto 3$. Thus, we have

$$\begin{pmatrix} 1 & 2 & 3 \\ 2 & 3 & 1 \end{pmatrix}^{-1} = \begin{pmatrix} 1 & 2 & 3 \\ 3 & 1 & 2 \end{pmatrix}.$$

Quiz 3.2 Write out all:

$$\begin{aligned} \begin{pmatrix} 1 & 2 & 3 \\ 1 & 2 & 3 \end{pmatrix} &= e, & \begin{pmatrix} 1 & 2 & 3 \\ 2 & 3 & 1 \end{pmatrix} &= (1\ 2\ 3), \\ \begin{pmatrix} 1 & 2 & 3 \\ 3 & 1 & 2 \end{pmatrix} &= (1\ 3\ 2), & \begin{pmatrix} 1 & 2 & 3 \\ 2 & 1 & 3 \end{pmatrix} &= (1\ 2), \\ \begin{pmatrix} 1 & 2 & 3 \\ 3 & 2 & 1 \end{pmatrix} &= (1\ 3), & \begin{pmatrix} 1 & 2 & 3 \\ 1 & 3 & 2 \end{pmatrix} &= (2\ 3). \end{aligned}$$

Quiz 3.3 Since $\sigma : 1 \mapsto 2,\ 2 \mapsto 3,\ 3 \mapsto 1$, we have $\sigma^2 : 1 \mapsto 3,\ 2 \mapsto 1,\ 3 \mapsto 2$, namely $\sigma^2 = (1\ 3\ 2)$.

We prove $\tau\sigma = \sigma^2\tau$.

$$\tau\sigma = \begin{pmatrix} 1 & 2 & 3 \\ 2 & 1 & 3 \end{pmatrix} \begin{pmatrix} 1 & 2 & 3 \\ 2 & 3 & 1 \end{pmatrix} = \begin{pmatrix} 1 & 2 & 3 \\ 1 & 3 & 2 \end{pmatrix} = (2\ 3),$$

$$\sigma^2\tau = \begin{pmatrix}1&2&3\\3&1&2\end{pmatrix}\begin{pmatrix}1&2&3\\2&1&3\end{pmatrix} = \begin{pmatrix}1&2&3\\1&3&2\end{pmatrix} = (2\ 3).$$

We prove $\tau\sigma^2 = \sigma\tau$.

$$\tau\sigma^2 = \begin{pmatrix}1&2&3\\2&1&3\end{pmatrix}\begin{pmatrix}1&2&3\\3&1&2\end{pmatrix} = \begin{pmatrix}1&2&3\\3&2&1\end{pmatrix} = (1\ 3),$$

$$\sigma\tau = \begin{pmatrix}1&2&3\\2&3&1\end{pmatrix}\begin{pmatrix}1&2&3\\2&1&3\end{pmatrix} = \begin{pmatrix}1&2&3\\3&2&1\end{pmatrix} = (1\ 3).$$

Quiz 3.4 The table for $\{e\}$ is trivial. The table for S_3 is already given. The other four tables are as follows:

	e	τ	Inverse	Order
e	e	τ	e	1
τ	τ	e	τ	2

	e	$\sigma\tau$	inverse	order
e	e	$\sigma\tau$	e	1
$\sigma\tau$	$\sigma\tau$	e	$\sigma\tau$	2

	e	$\sigma^2\tau$	Inverse	Order
e	e	$\sigma^2\tau$	e	1
$\sigma^2\tau$	$\sigma^2\tau$	e	$\sigma^2\tau$	2

	e	σ	σ^2	Inverse	Order
e	e	σ	σ^2	e	1
σ	σ	σ^2	e	σ^2	3
σ^2	σ^2	e	σ	σ	3

Quiz 3.5 The union is given by

$$\{e, \tau\} \cup \{e, \sigma, \sigma^2\} = \{e, \tau, \sigma, \sigma^2\},$$

in which neither $\tau\sigma$ nor $\sigma\tau$ can be found.

Quiz 3.6 The left and right A orbits passing through e coincide, in fact A itself. The left and right A orbits passing through τ are given by

$$A\tau = \{\tau, \sigma\tau, \sigma^2\tau\}, \quad \tau A = \{\tau, \sigma^2\tau, \sigma\tau\}.$$

We see that $A\tau = \tau A$.

Quiz 3.7 $Be = eB = B$ is trivial. The left and right B-orbits passing through σ are given by

$$B\sigma = \{\sigma, \sigma^2\tau\}, \quad \sigma B = \{\sigma, \sigma\tau\},$$

which show $B\sigma \neq \sigma B$.

The left and right B-orbits passing through σ^2 are given by

$$B\sigma^2 = \{\sigma^2, \sigma\tau\}, \quad \sigma^2 B = \{\sigma^2, \sigma^2\tau\},$$

which show $B\sigma^2 \neq \sigma^2 B$.

Quiz 3.8 The left and right B_1-orbits passing through σ are given by

$$B_1\sigma = \{\sigma, \tau\}, \quad \sigma B_1 = \{\sigma, \sigma^2\tau\},$$

which do not coincide.

The left and right B_2-orbits passing through σ are given by

$$B_2\sigma = \{\sigma, \sigma\tau\}, \quad \sigma B_2 = \{\sigma, \tau\},$$

which do not coincide.

Quiz 3.9 If we enumerate the powers of a^5, then we have all elements of C_6 as follows:

$$(a^5)^1 = a^5, \ (a^5)^2 = a^4, \ (a^5)^3 = a^3, \ (a^5)^4 = a^2, \ (a^5)^5 = a, \ (a^5)^6 = e.$$

Quiz 3.10 The order of an element of S_3 is 1, 2, or 3. So, no element can generate the whole S_3.

Quiz 3.11 We have already known that the set $\{\sigma, \tau\} =$ generates S_3.

$\{(1\ 2), (1\ 3)\}$ has $\tau = (1\ 2)$ as a member. We have $(1\ 3)(1\ 2) = (1\ 2\ 3) = \sigma$. Therefore, $\{(1\ 2), (1\ 3)\}$ generates S_3.

Quiz 3.12

	a^j	$a^j b$	inverse	order
a^i	a^{i+j}	$a^{i+j}b$	a^{6-i}	$\varphi(i)$
$a^i b$	$a^{i-j}b$	a^{i-j}	$a^i b$	2

The exponents of a are calculated modulo 6. So, we can safely write a^{-i} instead of a^{6-i}. $\varphi(i)$ is given by

$$\varphi(0) = 1, \ \varphi(1) = 6, \ \varphi(2) = 3, \ \varphi(3) = 2, \ \varphi(4) = 3, \ \varphi(5) = 6.$$

Quiz 3.13 For any n, the element $a^n b$ has order 2. They all generate subgroups of size 2.

Quiz 3.14

	e	a^3	b	a^3b	inverse	order
e	e	a^3	b	a^3b	e	1
a^3	a^3	e	a^3b	b	a^3	2
b	b	a^3b	e	a^3	b	2
a^3b	a^3b	b	a^3	e	a^3b	2

	e	a^3	ab	a^4b	inverse	order
e	e	a^3	ab	a^4b	e	1
a^3	a^3	e	a^4b	ab	a^3	2
ab	ab	a^4b	e	a^3	ab	2
a^4b	a^4b	ab	a^3	e	a^4b	2

	e	a^3	a^2b	a^5b	inverse	order
e	e	a^3	a^2b	a^5b	e	1
a^3	a^3	e	a^5b	a^2b	a^3	2
a^2b	a^2b	a^5b	e	a^3	a^2b	2
a^5b	a^5b	a^2b	a^3	e	a^5b	2

Quiz 3.15

$$\begin{aligned}
\{e, ab\}a &= \{a, b\} \neq \{a, a^2b\} = a\{e, ab\},\\
\{e, a^2b\}a &= \{a, ab\} \neq \{a, a^3b\} = a\{e, a^2b\},\\
\{e, a^3b\}a &= \{a, a^2b\} \neq \{a, a^4b\} = a\{e, a^3b\},\\
\{e, a^4b\}a &= \{a, a^3b\} \neq \{a, a^5b\} = a\{e, a^4b\},\\
\{e, a^5b\}a &= \{a, a^4b\} \neq \{a, b\} = a\{e, a^5b\}.
\end{aligned}$$

Quiz 3.16

$$\begin{aligned}
\{e, a^3, ab, a^4b\}a &= \{a, a^4, b, a^3b\} \neq \{a, a^4, a^2b, a^5b\} = a\{e, a^3, ab, a^4b\},\\
\{e, a^3, a^2b, a^5b\}a &= \{a, a^4, ab, a^4b\} \neq \{a, a^4, a^3b, b\} = a\{e, a^3, a^2b, a^5b\}.
\end{aligned}$$

Exercises

[1]

(1)

	e	a	a^2	a^3	a^4	a^5	a^6	a^7	a^8	a^9	a^{10}	a^{11}
order	1	12	6	4	3	12	2	12	3	4	6	12

(2) The elements of order 12 are a, a^5, a^7, and a^{11}. Each of them is a generator of C_{12}.

(3) We introduce the notation $\langle x\rangle = \{e, x^{\pm 1}, x^{\pm 2}, \cdots\}$. If x has a finite order n, then we have $\langle x\rangle = \{e, x, x^2, \cdots, x^{n-1}\}$.

$$\langle e\rangle = \{e\},$$

$$\langle a \rangle = \langle a^5 \rangle = \langle a^7 \rangle = \langle a^{11} \rangle = C_{12},$$
$$A = \langle a^2 \rangle = \langle a^{10} \rangle = \{e, a^2, a^4, a^6, a^8, a^{10}\},$$
$$B = \langle a^3 \rangle = \langle a^9 \rangle = \{e, a^3, a^6, a^9\},$$
$$C = \langle a^4 \rangle = \langle a^8 \rangle = \{e, a^4, a^8\},$$
$$D = \langle a^6 \rangle = \{e, a^6\}.$$

(4) Assume that $a^n \in N$ and that k does not divide n. By Euclidean division, there are natural numbers m and r that satisfy

$$n = km + r \quad (0 < r < k).$$

Since $a^k \in N$, we have $a^{km} \in N$, therefore $a^{-km} \in N$. Then, we have

$$a^n a^{-km} = a^r \in N,$$

which contradicts the minimality of k.

(5) By (4), the subgroups of C_{12} are all cyclic. Therefore, by (3), A, B, C, and D exhaust the non-trivial subgroups of C_{12}. The sets of orbits are given as follows.

$$C_{12}/A = \{A, aA\},$$
$$C_{12}/B = \{B, aB, a^2B\},$$
$$C_{12}/C = \{C, aC, a^2C, a^3C\},$$
$$C_{12}/D = \{D, aD, \cdots, a^5D\}.$$

[2]

(1) Assume that $a^k (|k| \geq 2)$ is a generator. Then, $a^{km} = a$ for some $|m| \geq 1$, therefore, $a^{km-1} = e$. Since $km - 1 \neq 0$, $a^{km-1} = e$ indicates the order of a is finite, which contradicts that a is a generator of the infinite cyclic group.

(2) Omitted.

(3) Any non-trivial subgroup of C_∞ is expressed as

$$\langle a^k \rangle = \{e, a^{\pm k}, a^{\pm 2k}, a^{\pm 3k}, \cdots\},$$

where $k \geq$ is a natural number.

[3]

(1) See Table A.3.

Table A.3 Table for D_4

Left \ Right	e	a	a^2	a^3	b	ab	a^2b	a^3b	Inverse	Order
e	e	a	a^2	a^3	b	ab	a^2b	a^3b	e	1
a	a	a^2	a^3	e	ab	a^2b	a^3b	b	a^3	4
a^2	a^2	a^3	e	a	a^2b	a^3b	b	ab	a^2	2
a^3	a^3	e	a	a^2	a^3b	b	ab	a^2b	a	4
b	b	a^3b	a^2b	ab	e	a^3	a^2	a	b	2
ab	ab	b	a^3b	a^2b	a	e	a^3	a^2	ab	2
a^2b	a^2b	ab	b	a^3b	a^2	a	e	a^3	a^2b	2
a^3b	a^3b	a^2b	ab	b	a^3	a^2	a	e	a^3b	2

The table shows the following. The set D_4 is closed under multiplication. It has the identity. Every element has its inverse. If we want a condensed formula similar to D_6 in Sect. 3.4, we have

$$a^i b^u \cdot a^j b^v = a^{i+(-1)^u j} b^{u+v},$$

where the exponents of a are calculated modulo 4 and exponents of b modulo 2. Associativity is automatic since D_4 has a geometric origin. If we prefer, we can calculate

$$(a^i b^u \cdot a^j b^v) \cdot a^k b^w = a^i b^u \cdot (a^j b^v \cdot a^k b^w)$$

by the above formula. Both sides should be

$$a^{i+(-1)^u j+(-1)^{u+v}k} b^{u+v+w}.$$

(2) Let us start with the subgroup $A_0 = \{e, a, a^2, a^3\}$ and its subgroups $A_1 = \{e, a^2\}$ and $\{e\}$. Let N be any other subgroup of D_4. Its subgroup $A_0 \cap N$ is also a subgroup of A_0. There are three possibilities:

$$A_0 \cap N = A_0, \quad A_1, \quad \text{or} \quad \{e\}.$$

(i) Assume $A_0 \cap N = A_0$.
Since $|N| > 4$ and $|N|$ divides $|D_4| = 8$, We have no option other than $N = D_4$.

(ii) Assume $A_0 \cap N = A_1$.
Take any elements $a^m b, a^n b \in N - A_0$. We have

$$(a^m b)(a^n b) = a^m (ba^n)b = a^m (a^{4-n} b)b = a^{m-n},$$

which means $a^{m-n} \in A_0 \cap N = A_1$. Therefore, $m - n \equiv 0 \mod 2$. We have two possibilities $N = B_1$ or $N = B_2$, where

$$B_1 = \{e, a^2, b, a^2b\} \quad \text{and} \quad B_2 = \{e, a^2, ab, a^3b\}.$$

(iii) Assume $A_0 \cap N = \{e\}$.
Take any elements $a^m b, a^n b \in N - A_0$. We have

$$(a^m b)(a^n b) = a^m (ba^n)b = a^m (a^{4-n} b)b = a^{m-n},$$

which means $a^{m-n} \in A_0 \cap N = \{e\}$. Therefore, $m = n$. The subgroup N is one of C_1, C_2, C_3, C_4 below:

$$C_1 = \{e, b\}, \quad C_2 = \{e, ab\}, \quad C_3 = \{e, a^2b\}, \quad C_4 = \{e, a^3b\}.$$

(3) Trivial subgroups $\{e\}$ and D_4 are normal subgroups. Any subgroup being half the size of its parent is a normal subgroup. Therefore, A_0, B_1, and B_2 are normal subgroups. Since $aC_1 = \{a, ab\}$ and $C_1a = \{a, a^3b\}$, the subgroup C_1 is not normal. Likewise, neither C_2, C_3, nor C_4 is normal. Since e and a^2 commute with all elements of D_4, the subgroup $A_1 = \{e, a^2\}$ is normal.

[4]

(1) The reader is recommended to work with a full table as in the previous exercise. However, the procedure is sheer repetition, so we omit the details. Alternatively, we present an approach based on a condensed table as follows:

	a^j	$a^j b$	inverse	order
a^i	a^{i+j}	$a^{i+j}b$	a^{5-i}	$\varphi(i)$
$a^i b$	$a^{i-j}b$	a^{i-j}	$a^i b$	2

Here, exponents of a are calculated modulo 5. So, we may write a^{-i} instead of a^{5-i}. The function φ is given by

$$\varphi(i) = \begin{cases} 1 & (\text{if } i = 0) \\ 5 & (\text{if } i \neq 0) \end{cases}.$$

It is easy to verify all conditions required to be a group. We can compactly express the multiplication rule as

$$a^i b^u \cdot a^j b^v = a^{i+(-1)^u j} b^{u+v},$$

where exponents of a and b are calculated modulo 5 and 2, respectively. Using this formula, we can prove associativity.

(2) Let us start with a subgroup $A_0 = \{e, a, a^2, a^3, a^4\}$. The group A_0 has only trivial subgroups $\{e\}$ and A_0. Let H be any other subgroup of D_5. Its subgroup $A_0 \cap H$ is also a subgroup of A_0. The possibilities are

$$A_0 \cap H = \{e\} \quad \text{or} \quad A_0.$$

(i) Assume $H \cap A_0 = \{e\}$.
Take any elements $a^i b, a^j b \in H - A_0$. We have $a^i b \cdot a^j b = a^{i-j} \in A_0 \cap H = \{e\}$. Thus, we have $i = j$. The subgroup H is one of

$$B_1{=}\{e, b\},\ \ B_2{=}\{e, ab\},\ \ B_3{=}\{e, a^2 b\},\ \ B_4{=}\{e, a^3 b\},\ \ B_5{=}\{e, a^4 b\},$$

all of which are cyclic groups of size 2.

(ii) Assume $H \cap A_0 = A_0$.
The only possibility is $H = D_5$.

(3) We have already known that $\{e\}$, A_0, and D_5 are normal subgroups of D_5. The subgroups B_i $(i = 1, \cdots, 5)$ are not normal. In fact, we have the following:

$$\{e, a^i b\}a = \{a, a^{i-1} b\},$$

$$a\{e, a^i b\} = \{a, a^{i+1} b\}.$$

These show that $B_i a \neq a B_i$ since $i - 1 \not\equiv i + 1 \mod 5$.

[5]

(1) We omit the table. Noting that 1 and -1 commute with all elements of Q_8 and that the whole system is symmetric among i, j, and k, we see that associativity is equivalent to a single equation $(ij)k = i(jk)$, which holds since both sides equal to -1.

(2) The cyclic subgroups generated by -1, $\pm i$, $\pm j$, and $\pm k$ are as follows:

$$\{1, -1\}, \quad \{\pm 1, \pm i\}, \quad \{\pm 1, \pm j\}, \quad \{\pm 1, \pm k\}.$$

Since there is only one element -1 of order 2, the subgroup of size 2 is $\{1, -1\}$ only. Other three subgroups above are cyclic of size 4.

Assume N is a subgroup of size 4 and not cyclic. Then, its elements have order 1 or 2. Such elements are limited to ± 1. Thus, this assumption yields nothing.

There remain trivial subgroups $\{1\}$ and Q_8.

(3) The trivial subgroups and those being half the size of the parent are all normal. We have to check if $\{1, -1\}$ is normal, which can be verified easily.

[6]

(1) The definition of matrix multiplication proves everything.
(2) Since $SL(n, \mathbb{C}) \subset GL(n, \mathbb{C})$, we have to prove that $SL(n, \mathbb{C})$ is closed under multiplication and inversion. Let $X, Y \in SL(n, \mathbb{C})$. Since $\det(XY) = \det X \times \det Y = 1$, the set is closed under multiplication. Since $\det(X^{-1}) = 1/\det X = 1$, the set is closed under inversion.
(3) Since $SL(n, \mathbb{Z}) \subset SL(n, \mathbb{C})$, only we have to do is to show that $SL(n, \mathbb{Z})$ is closed under multiplication and inversion. Let $X, Y \in SL(n, \mathbb{Z})$. The entries of XY are results of addition and multiplication. Therefore, they are all belong to $\mathbb{Z}$. We have $\det(XY) = \det X \times \det Y = 1$, which shows closedness under multiplication. We denote by $\tilde{X}$ the cofactor matrix of X. We know that

$$X^{-1} = \frac{1}{\det X}\tilde{X} = \tilde{X},$$

where entries of $\tilde{X}$ are results of addition and multiplication. Therefore, the entries of $\tilde{X}$ are all integers. Since $\det X^{-1} \times \det X = 1$ and $\det X = 1$, we have $\det X^{-1} = 1$. This shows that $SL(n, \mathbb{Z})$ is closed under inversion.

Chapter 4

Quiz 4.1 Let x has order k. Assume k does not divide n. Euclidean division gives natural number r such that $n = qk + r$ and $0 < r < k$. Then, we have

$$e = x^n = x^{qk+r} = (x^k)^q x^r = x^r,$$

which contradicts the minimality of k as the order of x.

Quiz 4.2 We first show that $\varphi(x(\ker\varphi)) = \{\varphi(x)\}$. For any $s \in \ker\varphi$, we have

$$\varphi(xs) = \varphi(x)\varphi(s) = \varphi(x)e_2 = \varphi(x),$$

which shows that the right $\ker\varphi$-orbit passing x is sent in cluster to $\{\varphi(x)\}$. Let $y(\ker\varphi)$ be another orbit. It is likewise sent in cluster to $\{\varphi(y)\}$. Suppose $\varphi(x) = \varphi(y)$. Then, we have

$$\varphi(x)\varphi(y^{-1}) = e_2.$$

The LHS is $\varphi(xy^{-1})$, therefore we have $xy^{-1} \in \ker\varphi$. This yields

$$x(\ker\varphi) = y(\ker\varphi),$$

namely the same right orbit.

Quiz 4.3

	e	τ	σ	$\sigma^2\tau$	σ^2	$\sigma\tau$
e	e	τ	σ	$\sigma^2\tau$	σ^2	$\sigma\tau$
τ	τ	e	$\sigma^2\tau$	σ	$\sigma\tau$	σ^2
σ	σ	$\sigma\tau$	σ^2	τ	e	$\sigma^2\tau$
$\sigma^2\tau$	$\sigma^2\tau$	σ^2	$\sigma\tau$	e	τ	σ
σ^2	σ^2	$\sigma^2\tau$	e	$\sigma\tau$	σ	τ
$\sigma\tau$	$\sigma\tau$	σ	τ	σ^2	$\sigma^2\tau$	e

Exercises

[1]

The image of a generator determines everything.

(1) Find homomorphisms $\varphi : C_2 \to C_4$.

The order of $\varphi(a)$ should be 1 or 2. Therefore, we have $\varphi(a) = e$ or $\varphi(a) = c^2$.

(i) Suppose $\varphi(a) = e$, the case of a trivial homomorphism.

$$\begin{array}{rccc} \varphi: & C_2 & \longrightarrow & C_4 \\ & e & \longmapsto & e \\ & a & \longmapsto & e \end{array}$$

$$\ker \varphi = C_2, \quad \operatorname{Im} \varphi = \{e\}.$$

(ii) Suppose $\varphi(a) = c^2$.

$$\begin{array}{rccc} \varphi: & C_2 & \longrightarrow & C_4 \\ & e & \longmapsto & e \\ & a & \longmapsto & c^2 \end{array}$$

$$\ker \varphi = \{e\}, \quad \operatorname{Im} \varphi = \{e, c^2\}.$$

(2) Find homomorphisms $\varphi : C_4 \to C_2$.

The order of $\varphi(c)$ should be divisors of 4, namely 1, 2, or 4. There is no element of order 4 in C_2. We have $\varphi(c) = e$ or $\varphi(c) = a$.

(i) Suppose $\varphi(c) = e$, the case of a trivial homomorphism.

$$\ker \varphi = C_4, \quad \operatorname{Im} \varphi = \{e\}.$$

(ii) Suppose $\varphi(c) = a$.

$$\begin{aligned} \varphi: \quad C_4 &\longrightarrow C_2 \\ e &\longmapsto e \\ c &\longmapsto a \\ c^2 &\longmapsto e \\ c^3 &\longmapsto a \end{aligned}$$

$$\ker \varphi = \{e, c^2\}, \quad \operatorname{Im} \varphi = C_2.$$

(3) Find homomorphisms $\varphi : C_3 \to C_6$.
The order of $\varphi(b)$ should be 1 or 3. The elements of order 3 in C_6 are c^2 and c^4. We have $\varphi(b) = e$, $\varphi(b) = d^2$, or $\varphi(b) = d^4$.

(i) Suppose $\varphi(b) = e$, the case of a trivial homomorphism.

$$\ker \varphi = C_3, \quad \operatorname{Im} \varphi = \{e\}.$$

(ii) Suppose $\varphi(b) = d^2$.

$$\begin{aligned} \varphi: \quad C_3 &\longrightarrow C_6 \\ e &\longmapsto e \\ b &\longmapsto d^2 \\ b^2 &\longmapsto d^4 \end{aligned}$$

$$\ker \varphi = \{e\}, \quad \operatorname{Im} \varphi = \{e, d^2, d^4\}.$$

(iii) Suppose $\varphi(b) = d^4$.

$$\begin{aligned} \varphi : C_3 &\longrightarrow C_6 \\ e &\longmapsto e \\ b &\longmapsto d^4 \\ b^2 &\longmapsto d^2 \end{aligned}$$

$$\ker \varphi = \{e\}, \quad \operatorname{Im} \varphi = \{e, d^2, d^4\}.$$

(4) Find homomorphisms $\varphi : C_3 \to C_4$.
The order of $\varphi(b)$ should be 1 or 3. There is no element of order 3 in C_4. There is no option other than $\varphi(b) = e$.

$$\ker \varphi = C_3, \quad \operatorname{Im} \varphi = \{e\}.$$

[2] Enumerate all excluding trivial homomorphisms.

(1)

$$\begin{array}{rrcl|rrcl|rrcl} \varphi : & C_2 & \longrightarrow & S_3 & \varphi : & C_2 & \longrightarrow & S_3 & \varphi : & C_2 & \longrightarrow & S_3 \\ & e & \longmapsto & e & & e & \longmapsto & e & & e & \longmapsto & e \\ & a & \longmapsto & \tau & & a & \longmapsto & \sigma\tau & & a & \longmapsto & \sigma^2\tau \end{array}$$

In all the cases, we have $\ker \varphi = \{e\}$ and obvious Im.

(2)

$$\begin{array}{rrcl|rrcl} \varphi : & C_3 & \longrightarrow & S_3 & \varphi : & C_3 & \longrightarrow & S_3 \\ & e & \longmapsto & e & & e & \longmapsto & e \\ & b & \longmapsto & \sigma & & b & \longmapsto & \sigma^2 \\ & b^2 & \longmapsto & \sigma^2 & & b^2 & \longmapsto & \sigma \end{array}$$

In both cases, we have $\ker \varphi = \{e\}$ and $\operatorname{Im} \varphi = \{e, \sigma, \sigma^2\}$.

(3) Find homomorphisms $\varphi : C_6 \to S_3$.
The order of $\varphi(d)$ is 1, 2, 3, or 6. There is no element of order 6 in S_3. The element $\varphi(d)$ is one of $e, \tau, \sigma\tau, \sigma^2\tau, \sigma$, or σ^2.
We present one generic example for each case where the order of $\varphi(d)$ is 2 or 3.

(i) Suppose $\varphi(d) = \sigma\tau$.

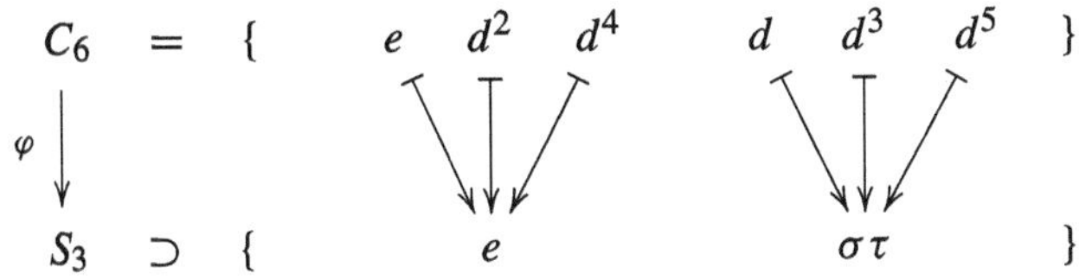

$$\ker\varphi = \{e, d^2, d^4\}, \quad \mathrm{Im}\,\varphi = \{e, \sigma\tau\}.$$

(ii) Suppose $\varphi(d) = \sigma$.

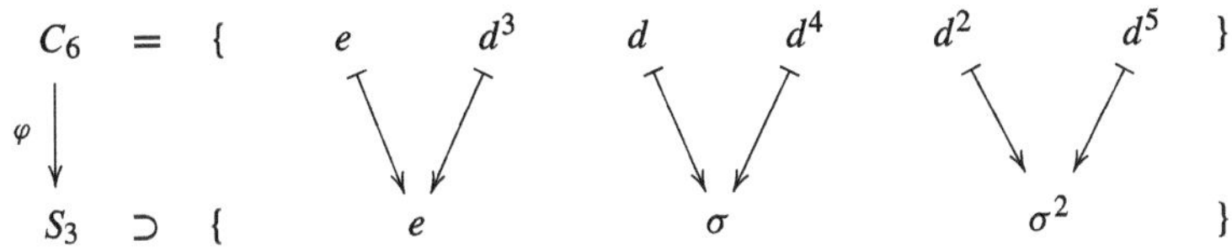

$$\ker\varphi = \{e, d^3\}, \quad \mathrm{Im}\,\varphi = \{e, \sigma, \sigma^2\}.$$

[3]

(1) Study factor groups of D_4. We omit cases of the trivial normal subgroups. The normal subgroups are as follows. See the solution to Exercise [3] of Chap. 3.

$$A_0 = \{e, a, a^2, a^3\},$$
$$A_1 = \{e, a^2\},$$
$$B_1 = \{e, a^2, b, a^2b\},$$
$$B_2 = \{e, a^2, ab, a^3b\}.$$

The factor groups by A_0, B_1, and B_2 are as follows:

$$D_4/A_0 = \{A_0, bA_0\},$$
$$D_4/B_1 = \{B_1, bB_1\},$$
$$D_4/B_2 = \{B_2, bB_2\}.$$

All these are isomorphic to C_2. We give Table A.4 for D_4/A_0 and omit others. The factor group by A_1, namely $D_4/A_1 = \{A_1, aA_1, bA_1, abA_2\}$, has Table A.5.

Table A.4 Multiplication in D_4/A_0

Left \ Right	A_0	bA_0	Inverse	Order
A_0	A_0	bA_0	A_0	1
bA_0	bA_0	A_0	bA_0	2

Table A.5 Multiplication in D_4/A_1

Left \ Right	A_1	aA_1	bA_1	abA_1	Inverse	Order
A_1	A_1	aA_1	bA_1	abA_1	A_1	1
aA_1	aA_1	A_1	abA_1	bA_1	aA_1	2
bA_1	bA_1	abA_1	A_1	aA_1	bA_1	2
abA_1	abA_1	bA_1	aA_1	A_1	abA_1	2

(2) Show the factor groups of Q_8 excluding the ones by trivial subgroups. The normal subgroups are $\{1, -1\}$, $\{\pm 1, \pm i\}$, $\{\pm 1, \pm j\}$, and $\{\pm 1, \pm k\}$.
Multiplication in

$$Q_8/\{1, -1\} = \{\{1, -1\}, \{i, -i\}, \{j, -j\}, \{k, -k\}\}$$

is given in Table A.6.

Table A.6 Multiplication in $Q_8/\{\pm 1\}$

Left \ Right	$\{\pm 1\}$	$\{\pm i\}$	$\{\pm j\}$	$\{\pm k\}$	Inverse	Order
$\{\pm 1\}$	$\{\pm 1\}$	$\{\pm i\}$	$\{\pm j\}$	$\{\pm k\}$	$\{\pm 1\}$	1
$\{\pm i\}$	$\{\pm i\}$	$\{\pm 1\}$	$\{\pm k\}$	$\{\pm j\}$	$\{\pm i\}$	2
$\{\pm j\}$	$\{\pm j\}$	$\{\pm k\}$	$\{\pm 1\}$	$\{\pm i\}$	$\{\pm j\}$	2
$\{\pm k\}$	$\{\pm k\}$	$\{\pm j\}$	$\{\pm i\}$	$\{\pm 1\}$	$\{\pm k\}$	2

Multiplication in

$$Q_8/\{\pm 1, \pm i\} = \{\{\pm 1, \pm i\}, \{\pm j, \pm k\}$$

is given in Table A.7. Other factor groups can be presented in the same way.

Table A.7 Multiplication in $Q_8/\{\pm 1, \pm i\}$

Left \ Right	$\{\pm 1, \pm i\}$	$\{\pm j, \pm k\}$	Inverse	Order
$\{\pm 1, \pm i\}$	$\{\pm 1, \pm i\}$	$\{\pm j, \pm k\}$	$\{\pm 1, \pm i\}$	1
$\{\pm j, \pm k\}$	$\{\pm j, \pm k\}$	$\{\pm 1, \pm i\}$	$\{\pm j, \pm k\}$	2

Chapter 5

Quiz 5.1 Let $x \in \{n_1, n_2, \cdots, n_\ell\}$. From the disjointness assumption we know $x \notin \{m_1, m_2, \cdots, m_k\}$, which means $(m_1 m_2 \cdots m_k)x = x$. Thus, we have

$$(n_1 n_2 \cdots n_\ell)(m_1 m_2 \cdots m_k)x = (n_1 n_2 \cdots n_\ell)x.$$

Since $(n_1 n_2 \cdots n_\ell)x \in \{n_1, n_2, \cdots, n_\ell\}$, disjointness gives $(n_1 n_2 \cdots n_\ell)x \notin \{m_1, m_2, \cdots, m_k\}$. Therefore, we have

$$(m_1 m_2 \cdots m_k)\underline{(n_1 n_2 \cdots n_\ell)x} = (n_1 n_2 \cdots n_\ell)x.$$

Two results coincide.

Quiz 5.2 Since $(1\ 2\ 3)(1\ 2) = (1\ 3)$, its third power is $(1\ 3)$. Since $(1\ 2\ 3)^3 = e$ and $(1\ 2)^3 = (1\ 2)$, we have $(1\ 2\ 3)^3(1\ 2)^3 = (1\ 2)$.

Quiz 5.3 We express a cycle with its smallest integer at the head. Cycles of length 3 that contain 1 are lexicographically listed below:

$$(1\ 2\ 3),\ (1\ 2\ 4),\ (1\ 3\ 2),$$
$$(1\ 3\ 4),\ (1\ 4\ 2),\ (1\ 4\ 3).$$

Those that do not contain 1 are listed below:

$$(2\ 3\ 4),\ (2\ 4\ 3).$$

We have a total of 8 cycles of length 3.

Quiz 5.4 Choosing the partner of 1 in a transposition determines a conjugate of $(1\ 3)(2\ 4)$. There are only two elements: $(1\ 2)(3\ 4)$ and $(1\ 4)(2\ 3)$.

Exercises

[1]

(1) $(1\ 5\ 7)(2\ 4)(3\ 6\ 8)$. The order is 6.
(2) $(1\ 5)(2\ 7\ 8\ 4\ 6\ 3)$. The order is 6.
(3) $(1\ 3\ 5)(4\ 7\ 8)$. The order is 3.

[2]

(1) Enumerate all transpositions in S_5. In lexicographical order, we have the following:

$$\begin{aligned}(1\ 2), (1\ 3), (1\ 4), (1\ 5),\\ (2\ 3), (2\ 4), (2\ 5),\\ (3\ 4), (3\ 5),\\ (4\ 5).\end{aligned}$$

(2) For each of the transpositions obtained in (1), we consider cycles of length 3 using the remaining three digits. The number of such cycles is 2 for each transposition. Thus, the number of conjugates is $2 \times 10 = 20$.
(3) Set a as follows:

$$a = \begin{pmatrix} 1\ 2\ 3\ 4\ 5 \\ 4\ 5\ 3\ 1\ 2 \end{pmatrix} = (1\ 4)(2\ 5).$$

The reader should verify that this a does its job.

Alternatively, set $a = (1\ 2\ 3\ 4\ 5)^3 = (1\ 4\ 2\ 5\ 3)$. Then, $a^{-1} = (1\ 2\ 3\ 4\ 5)^2 = (1\ 3\ 5\ 2\ 4)$. The alternative a also does its job.

We can systematically study the freedom of choice of a using the notions of *normalizer* and *centralizer*. Consult with standard textbooks.

[3]

The reader may find the solution below too lengthy. This exercise is intended to give the reader to have an intensive experience to apply basic techniques in combination through tackling a concrete task.

(1) Groups of size 2 are cyclic groups. Subgroups of size 2 are in one-to-one correspondence with elements of order 2 as generators. In S_4, elements of order 2 are conjugate to either $(1\ 2)$ or $(1\ 2)(3\ 4)$.

The number of elements conjugate to $(1\ 2)$ is 6. The following 6 subgroups are conjugate with each other:

$$\{e, (1\ 2)\}, \{e, (1\ 3)\}, \{e, (1\ 4)\}, \{e, (2\ 3)\}, \{e, (2\ 4)\}, \{e, (3\ 4)\}.$$

The number of elements conjugate to $(1\ 2)(3\ 4)$ is 3. The following 3 subgroups are conjugate to each other:

$$\{e, (1\ 2)(3\ 4)\}, \quad \{e, (1\ 3)(2\ 4)\}, \quad \{e, (1\ 4)(2\ 3)\}.$$

(2) Groups of size 3 are cyclic groups. Subgroups of size 3 are in one-to-TWO correspondence with elements of order 3 as generators. Why one-to-two? The square of a generator of a subgroup of size 3 is an alternative generator of the same subgroup. Elements of order 3 in S_4 are cycles of length 3. The total number of them is 8. So, there are 4 subgroups of size 3 in S_4 as follows:

$$\{e, (1\ 2\ 3), (1\ 3\ 2)\}, \quad \{e, (1\ 2\ 4), (1\ 4\ 2)\},$$

$$\{e, (1\ 3\ 4), (1\ 4\ 3)\}, \quad \{e, (2\ 3\ 4), (2\ 4\ 3)\}.$$

All of them are conjugates to each other.

(3) Groups of size 4 may or may not be cyclic.

Cyclic subgroups of size 4 are in one-to-two correspondence with elements of order 4 as generators. The third power of a generator of a subgroup of size 4 is an alternative generator of the same subgroup. Elements of order 4 in S_4 are cycles of length 4. The total number of them is 6. So, there are 3 cyclic subgroups of size 4 in S_4 as follows:

$$\{e, (1\ 2\ 3\ 4), (1\ 3)(2\ 4), (1\ 4\ 3\ 2)\},$$
$$\{e, (1\ 2\ 4\ 3), (1\ 4)(2\ 3), (1\ 3\ 4\ 2)\},$$
$$\{e, (1\ 3\ 2\ 4), (1\ 2)(3\ 4), (1\ 4\ 2\ 3)\}.$$

All of them are conjugates to each other.

Non-cyclic subgroups of size 4 do not contain elements of order 4 or 3. Such a subgroup consists of three elements of order 2 except for the identity. We have already seen the catalog of elements of order 2 in (1).

Let K be a non-cyclic subgroup of size 4. Let $a \neq b \in K$ be elements of order 2. We are forced to have $ab = ba$. In fact, if $ab \neq ba$, then $\{e, a, b, ab, ba\} \subset K$, the LHS has size 5. This contradicts the assumption. So, we have $K = \{e, a, b, ab\} \subset K$ with $ab = ba$ and $a^2 = b^2 = (ab)^2 = e$.

We only have to choose commutative a and b from the set of elements of order 2. If a and b are both transpositions, then the possibilities are as follows:

$$\{e, (1\ 2), (3\ 4), (1\ 2)(3\ 4)\}, \qquad \{e, (1\ 3), (2\ 4), (1\ 3)(2\ 4)\},$$
$$\{e, (1\ 4), (2\ 3), (1\ 4)(2\ 3)\}.$$

If we allow b to be the product of a pair of disjoint transpositions, the results are already included in the three cases above. All of these are conjugate to each other.

What remains to be checked is the case where $K = \{e, a, b, c\}$ where a, b, c are products of disjoint transpositions. The only possibility left is as follows:

$$\{e, \quad (1\ 2)(3\ 4), \quad (1\ 3)(2\ 4), \quad (1\ 4)(2\ 3)\}.$$

This is a normal subgroup of S_4.

(4) There is no cyclic subgroup of size 6 in S_4 since there is no element of order 6 in S_4. All subgroups of size 6 should be generated by elements of order 2 and 3.

A subgroup generated by a cycle a of length 3 and a transposition b consisting of numbers in a, say $a = (2\ 3\ 4)$ and $b = (2\ 3)$ is a group isomorphic to S_3. The same subgroup is obtained if we replace $b = (2\ 3)$ with $b = (2\ 4)$ or $b = (3\ 4)$. The number of such conjugate subgroups is 4.

A subgroup generated by a cycle a of length 3 and a transposition b including a number not in a, say $a = (1\ 2\ 3)$ and $b = (3\ 4)$. We have

$$(1\ 2\ 3)(3\ 4) = (1\ 2\ 3\ 4).$$

We have that ab is a cycle of length 4. In general, we have

$$\begin{aligned} &(b\ c\ a)(c\ d) \\ =\ &(c\ a\ b)(c\ d) \\ =\ &(a\ b\ c)(c\ d) \\ =\ &(a\ b\ c\ d). \end{aligned}$$

So, this case should be abandoned. Thus, we can conclude that the number of subgroups of size 6 in S_4 is 4, all of which are conjugate to each other and isomorphic to S_3.

(5) (i) There is no cyclic subgroup of size 8 in S_4 since there is no element of order 8 in S_4.

(ii) We seek a subgroup, say H, all of whose elements have order 2 except for the identity.

There are 9 elements of order 2 in S_4: 6 transpositions and 3 even permutations. Even permutations are products of disjoint transpositions as follows:

$$(1\ 2)(3\ 4), \quad (1\ 3)(2\ 4), \quad (1\ 4)(2\ 3).$$

The product of any two commuting elements of order 2 has order 2. What happens when non-commuting elements of order 2 are multiplied? There are only two patterns for such multiplication:

* $(a\ b) \cdot (b\ c) = (a\ b\ c)$ yields a cycle of length 3.
* $(a\ c) \cdot (a\ b)(c\ d) = (a\ b\ c\ d)$ or $(a\ b)(c\ d) \cdot (a\ c) = (a\ d\ b\ c)$. In either case, there arises a cycle of length 4.

Here $\{a, b, c, d\} = \{1, 2, 3, 4\}$. To attain the goal we must avoid non-commuting elements.

Assume $(1\ 2) \in H$. Elements of order 2 that do not commute with $(1\ 2)$ are enumerated as follows:

$$(1\ 3),\ (1\ 4),\ (2\ 3),\ (2\ 4),\ (1\ 3)(2\ 4),\ (1\ 4)(2\ 3).$$

There are in total 6, and these must be excluded from H. Since $9 - 6 + 1 < 8$, we see $(1\ 2) \notin H$. By symmetry, we conclude that the subgroup H should not contain transpositions.

The remaining elements of order 2 are even permutations the number of which is 3. So, we cannot construct H out of these.

(iii) The above arguments show that a subgroup of size 8, say H, in S_4 should be generated by elements of order 2 and 4.

Assume $(1\ 2) \in H$. Its product with cycles of length 4 are as follows:

$$\begin{aligned}
(1\ 2\ 3\ 4)(1\ 2) &= (1\ 3\ 4),\\
(1\ 2\ 4\ 3)(1\ 2) &= (1\ 4\ 3),\\
(1\ 3\ 2\ 4)(1\ 2) &= (1\ 4)(2\ 3),\\
(1\ 3\ 4\ 2)(1\ 2) &= (2\ 3\ 4),\\
(1\ 4\ 2\ 3)(1\ 2) &= (1\ 3)(2\ 4),\\
(1\ 4\ 3\ 2)(1\ 2) &= (2\ 4\ 3).
\end{aligned}$$

The subgroup H should avoid cycles of length 3. Thus, candidates for members are limited to

$$(1\ 3\ 2\ 4) \quad \text{and} \quad (1\ 4\ 2\ 3),$$

each of which is the third power of the other. Thus, both of these are members of H or neither of them. If they are members, we should test the following:

$$(1\ 3)(2\ 4) \cdot (1\ 4)(2\ 3) = (1\ 2)(3\ 4).$$

We see $(1\ 3\ 2\ 4)^2 = (1\ 4\ 2\ 3)^2$ and $(1\ 2)\ (1\ 2)(3\ 4) = (3\ 4)$. So far, we have a total of 8 elements. Making the following table is a good try.

	e	(1324)	(12)(34)	(1423)	(12)	(34)	(13)(24)	(14)(23)
e	e	(1324)	(12)(34)	(1423)	(12)	(34)	(13)(24)	(14)(23)
(1324)	(1324)	(12)(34)	(1423)	e	(14)(23)	(13)(24)	(12)	(34)
(12)(34)	(12)(34)	(1423)	e	(1324)	(34)	(12)	(14)(23)	(13)(24)
(1423)	(1423)	e	(1324)	(12)(34)	(13)(24)	(14)(23)	(34)	(12)
(12)	(12)	(13)(24)	(34)	(14)(23)	e	(12)(34)	(1324)	(1423)
(34)	(34)	(14)(23)	(12)	(13)(24)	(12)(34)	e	(1423)	(1324)
(13)(24)	(13)(24)	(34)	(14)(23)	(12)	(1423)	(1324)	e	(12)(34)
(14)(23)	(14)(23)	(12)	(13)(24)	(34)	(1324)	(1423)	(12)(34)	e

This table shows that the set forms a subgroup of S_4.

We can start the same argument with $(3\ 4)$ to get to the same subgroup. The conjugate subgroups including this are subgroups:

* generated by $(1\ 2)$ and $(1\ 3\ 2\ 4)$,

  * generated by (1 3) and (1 2 3 4),
  * generated by (1 4) and (1 3 4 2).

(6) Subgroups of size 12 are all normal subgroups of S_4 since they are half the size of S_4. Let H be such a normal subgroup. If $x \in H$, then the conjugate elements of x must all be contained in H. The conjugate classes are enumerated as follows:

- cycles of length 4. There are 6.
- cycles of length 3. There are 8.
- products of disjoint transpositions. There are 3.
- transpositions. There are 6.
- the identity. There is 1.

To get the sum 12 including 1, the only addition is $8 + 3 + 1$. We can conclude that the only subgroup of size 12 in S_4 is the alternating group A_4 consisting of all even permutations.

Chapter 6

Quiz 6.1 $H \cap N$ is obvious. The following table gives HN.

$H \downarrow \quad N \rightarrow$	e	(1 2)(3 4)	(1 3)(2 4)	(1 4)(2 3)
e	e	(1 2)(3 4)	(1 3)(2 4)	(1 4)(2 3)
(1 2 3 4)	(1 2 3 4)	(1 3)	(1 4 3 2)	(2 4)
(1 3)(2 4)	(1 3)(2 4)	(1 4)(2 3)	e	(1 2)(3 4)
(1 4 3 2)	(1 4 3 2)	(2 4)	(1 2 3 4)	(1 3)

Quiz 6.2 Since $gNg^{-1} = N$ for all $g \in G$, we have hNh^{-1} for all $h \in H \subset G$.

Exercises

[1]

(1) The normal subgroups of S_3 are $\{e\}$, $\{e, \sigma, \sigma^2\}$, and S_3. Let $\varphi : S_3 \to C_6 = \{e, a, \cdots, a^5\}$ be a homomorphism.

(i) Assume $\ker\varphi = \{e\}$.
Since φ is injective and $|S_3| = |C_6| < \infty$, the map φ is bijective. There is an element a of order 6 in C_6. The order of $\varphi^{-1}(a)$ must be a multiple of 6, which is impossible.

(ii) Assume $\ker\varphi = \{e, \sigma, \sigma^2\}$.
The homomorphism φ is determined as follows:

$$m \cdot (n \cdot x) = \underbrace{\underbrace{(x + \cdots + x)}_{n \text{ times}} + \cdots + \underbrace{(x + \cdots + x)}_{n \text{ times}}}_{m \text{ times}} = \underbrace{x + \cdots + x}_{mn \text{ times}} = (mn) \cdot x.$$

Quiz 9.3 The set N_1 is closed under scalar multiplication:

$$\begin{pmatrix} a_1 & 0 \\ 0 & a_2 \end{pmatrix} \begin{pmatrix} z \\ 0 \end{pmatrix} = \begin{pmatrix} a_1 z \\ 0 \end{pmatrix}.$$

So is the set N_2.

Quiz 9.4 The set N_1 is closed under A-scalar multiplication:

$$\begin{pmatrix} a_{11} & a_{12} & 0 \\ a_{21} & a_{22} & 0 \\ 0 & 0 & a_{33} \end{pmatrix} \begin{pmatrix} x \\ y \\ 0 \end{pmatrix} = \begin{pmatrix} a_{11}x + a_{12}y \\ a_{21}x + a_{22}y \\ 0 \end{pmatrix}.$$

The set N_2 is closed under A-scalar multiplication:

$$\begin{pmatrix} a_{11} & a_{12} & 0 \\ a_{21} & a_{22} & 0 \\ 0 & 0 & a_{33} \end{pmatrix} \begin{pmatrix} 0 \\ 0 \\ z \end{pmatrix} = \begin{pmatrix} 0 \\ 0 \\ a_{33}z \end{pmatrix}.$$

Quiz 9.5 For any $\begin{pmatrix} x \\ y \\ 0 \end{pmatrix} \in N_1$, we seek a 2×2 matrix $P = \begin{pmatrix} a_{11} & a_{12} \\ a_{21} & a_{22} \end{pmatrix}$ such that

$$\begin{pmatrix} x \\ y \\ 0 \end{pmatrix} = \begin{pmatrix} a_{11} & a_{12} & 0 \\ a_{21} & a_{22} & 0 \\ 0 & 0 & a_{33} \end{pmatrix} \begin{pmatrix} a \\ b \\ 0 \end{pmatrix}.$$

We can ignore a_{33} and continue calculating with 2×2 matrices. Let P be a matrix satisfying

$$P \begin{pmatrix} a \\ b \end{pmatrix} = \begin{pmatrix} x \\ y \end{pmatrix}, \quad P \begin{pmatrix} -\bar{b} \\ \bar{a} \end{pmatrix} = \begin{pmatrix} 0 \\ 0 \end{pmatrix}.$$

Using the techniques in Sect. 1.5, we have

$$P \begin{pmatrix} a & -\bar{b} \\ b & \bar{a} \end{pmatrix} = \begin{pmatrix} x & 0 \\ y & 0 \end{pmatrix}.$$

Solving this equation with the unknown matrix P, we have

$$P = \begin{pmatrix} x & 0 \\ y & 0 \end{pmatrix} \begin{pmatrix} a & -\bar{b} \\ b & \bar{a} \end{pmatrix}^{-1} = \frac{1}{|a|^2 + |b|^2} \begin{pmatrix} \bar{a}x & \bar{b}x \\ \bar{a}y & \bar{b}y \end{pmatrix}.$$

This shows that N_1 is the A-submodule generated by $\begin{pmatrix} x \\ y \\ 0 \end{pmatrix} \in N_1$.

Similar but much simpler arguments can be done on N_2.

Quiz 9.6 It is sufficient to find a counter-example to

$$\begin{pmatrix} 0 & 1 \\ 1 & 0 \end{pmatrix} \begin{pmatrix} a_1 & 0 \\ 0 & a_2 \end{pmatrix} = \begin{pmatrix} a_1 & 0 \\ 0 & a_2 \end{pmatrix} \begin{pmatrix} 0 & 1 \\ 1 & 0 \end{pmatrix}$$

to show that φ is not left A-module homomorphism. The LHS is $\begin{pmatrix} 0 & a_2 \\ a_1 & 0 \end{pmatrix}$, while the RHS is $\begin{pmatrix} 0 & a_1 \\ a_2 & 0 \end{pmatrix}$. When $a_1 \neq a_2$, the equality does not hold.

The map ψ is left A-module homomorphism since

$$\begin{pmatrix} 2 & 0 \\ 0 & 3 \end{pmatrix} \begin{pmatrix} a_1 & 0 \\ 0 & a_2 \end{pmatrix} = \begin{pmatrix} a_1 & 0 \\ 0 & a_2 \end{pmatrix} \begin{pmatrix} 2 & 0 \\ 0 & 3 \end{pmatrix}$$

is always true.

Quiz 9.7 We prove (1). The LHS is

$$(a_1 + a_2) \bullet (x + L) = (a_1 + a_2)x + L = (a_1x + a_2x) + L,$$

while the RHS is

$$a_1 \bullet (x + L) + a_2 \bullet (x + L) = (a_1x + L) + (a_2x + L) = (a_1x + a_2x) + L,$$

which shows that they are equal.

We prove (2). The LHS is

$$a \bullet ((x_1 + L) + (x_2 + L)) = a \bullet ((x_1 + x_2) + L) = a(x_1 + x_2) + L = (ax_1 + ax_2) + L,$$

while the RHS is

$$a \bullet (x_1 + L) + a \bullet (x_2 + L) = (ax_1 + L) + (ax_2 + L) = (ax_1 + ax_2) + L,$$

which shows that they are equal.

We prove (3). The LHS is

$$(a_1a_2) \bullet (x + L) = a_1a_2x + L,$$

while the RHS is

$$a_1 \bullet (a_2 \bullet (x + L)) = a_1 \bullet (a_2 x + L) = a_1 a_2 x + L,$$

which shows that they are equal.

Condition (4) is trivial.

Quiz 9.8 We first show that $X\,k[X]$ is a subgroup of addition. For any $Xf(X), Xg(X) \in X\,[X]$, we have

$$Xf(X) + Xg(X) = X(f(X) + g(X)) \in X\,k[X],$$

which means $X\,k[X]$ is closed under addition. The additive inverse of $Xf(X)$ is $X(-f(X))$. The neutral element is $X \times 0 = 0$.

Next, we show that $X\,k[X]$ is closed under scalar multiplication. Let $f(X) \in k[X]$ and $Xg(X) \in X\,k[X]$ be any scalar and an element of $X\,k[X]$. We have

$$f(X) \bullet Xg(X) = Xf(X)g(X) \in X\,k[X].$$

We conclude that $X\,k[X]$ is a $k[X]$-submodule.

We show that φ is surjective. For any $a \in k$, we take a polynomial $f(X) \in k[X]$ such that $f(0) = a$, for examples $f(X) = a$ or $f(X) = X^2 + X + a$. We have

$$\varphi(f(X)) = f(0) = a.$$

Quiz 9.9 We first prove $W = U + V_2$. It is clear that $W \supset U + V_2$. For any $\begin{pmatrix} x \\ y \end{pmatrix}$, we have

$$\begin{pmatrix} x \\ y \end{pmatrix} = \begin{pmatrix} x - y \\ 0 \end{pmatrix} + \begin{pmatrix} y \\ y \end{pmatrix} \in U + V_2,$$

which shows that $W \subset U + V_2$.

If we set $\begin{pmatrix} x \\ 0 \end{pmatrix} = \begin{pmatrix} y \\ y \end{pmatrix}$, then we have $x = y = 0$, which shows that $U \cap V_2 = \left\{\begin{pmatrix} 0 \\ 0 \end{pmatrix}\right\}$.

Quiz 9.10 The following calculations complete the proofs that $2C_6$ and $3C_6$ are C_6-submodules:

$4\times$			$5\times$		
$2C_6$	$3C_6$	$2C_6+3C_6$	$2C_6$	$3C_6$	$2C_6+3C_6$
0	0	0	0	0	0
0	0	0	0	3	3
2	0	2	4	0	4
2	0	2	4	3	1
4	0	4	2	0	2
4	0	4	2	3	5

The scalar multiplication $4\times$ is the projection $2C_6 \oplus 3C_6 \to 2C_6$.

Quiz 9.11 Scalar multiplications are given below.

For L_0, we have

$$\begin{pmatrix} a & b \\ 0 & c \end{pmatrix} \begin{pmatrix} x & 0 \\ 0 & 0 \end{pmatrix} = \begin{pmatrix} ax & 0 \\ 0 & 0 \end{pmatrix} \in L_0.$$

For L_∞, we have

$$\begin{pmatrix} a & b \\ 0 & c \end{pmatrix} \begin{pmatrix} 0 & y \\ 0 & z \end{pmatrix} = \begin{pmatrix} 0 & ay+bz \\ 0 & cz \end{pmatrix} \in L_\infty.$$

For N, we have

$$\begin{pmatrix} a & b \\ 0 & c \end{pmatrix} \begin{pmatrix} 0 & y \\ 0 & 0 \end{pmatrix} = \begin{pmatrix} 0 & ay \\ 0 & 0 \end{pmatrix} \in N.$$

Exercises

[1] Straightforward calculation will complete the task. We have an alternative approach. Let the $\mathbb{C}$-module structure, namely the ring homomorphism $\mathbb{C} \to \mathrm{End}(\mathbb{R}^2) = M_2(\mathbb{R})$, be expressed as

$$x + y\sqrt{-1} \mapsto x \begin{pmatrix} 1 & 0 \\ 0 & 1 \end{pmatrix} + y \begin{pmatrix} 0 & -1 \\ 1 & 0 \end{pmatrix}.$$

(1) The composition of the scalar multiplications $(x + y\sqrt{-1})\bullet$ and $(u + v\sqrt{-1})\bullet$ is given in the matrix form as

$$\left(x \begin{pmatrix} 1 & 0 \\ 0 & 1 \end{pmatrix} + y \begin{pmatrix} 0 & -1 \\ 1 & 0 \end{pmatrix}\right) \left(u \begin{pmatrix} 1 & 0 \\ 0 & 1 \end{pmatrix} + v \begin{pmatrix} 0 & -1 \\ 1 & 0 \end{pmatrix}\right).$$

Noting that

$$\begin{pmatrix} 0 & -1 \\ 1 & 0 \end{pmatrix}^2 = -\begin{pmatrix} 1 & 0 \\ 0 & 1 \end{pmatrix},$$

we can easily obtain

$$(xy - yv)\begin{pmatrix} 1 & 0 \\ 0 & 1 \end{pmatrix} + (xv + yu)\begin{pmatrix} 0 & -1 \\ 1 & 0 \end{pmatrix}.$$

Since $(x + y\sqrt{-1})(u + v\sqrt{-1}) = (xy - yv) + (xv + yu)\sqrt{-1}$, we have the desired equation.

(2) Omitting the column vector $\begin{pmatrix} a \\ b \end{pmatrix}$ from the both sides, the RHS is

$$\begin{aligned}
&\left(x\begin{pmatrix} 1 & 0 \\ 0 & 1 \end{pmatrix} + y\begin{pmatrix} 0 & -1 \\ 1 & 0 \end{pmatrix}\right) + \left(u\begin{pmatrix} 1 & 0 \\ 0 & 1 \end{pmatrix} + v\begin{pmatrix} 0 & -1 \\ 1 & 0 \end{pmatrix}\right) \\
=&(x + u)\begin{pmatrix} 1 & 0 \\ 0 & 1 \end{pmatrix} + (y + v)\begin{pmatrix} 0 & -1 \\ 1 & 0 \end{pmatrix},
\end{aligned}$$

while the LHS is the result of

$$\begin{aligned}
&(x + y\sqrt{-1}) + (u + v\sqrt{-1}) \\
&= (x + u) + (y + v)\sqrt{-1} \mapsto (x + u)\begin{pmatrix} 1 & 0 \\ 0 & 1 \end{pmatrix} + (y + v)\begin{pmatrix} 0 & -1 \\ 1 & 0 \end{pmatrix}.
\end{aligned}$$

(3) Omitted.
(4) Omitted.

[2] Straightforward calculation will complete the task. An alternative approach like [1] is to define the ring homomorphism $\mathbb{C} \to \mathrm{End}(\mathbb{R}^2) = M_2(\mathbb{R})$ by

$$x + y\sqrt{-1} \mapsto x\begin{pmatrix} 1 & 0 \\ 0 & 1 \end{pmatrix} + y\begin{pmatrix} 1 & -2 \\ 1 & -1 \end{pmatrix}.$$

Use the following equation:

$$\begin{pmatrix} 1 & -2 \\ 1 & -1 \end{pmatrix}^2 = -\begin{pmatrix} 1 & 0 \\ 0 & 1 \end{pmatrix}.$$

[3] We can focus on the 2×2 block in

$$\begin{pmatrix} * & * & \\ * & * & \\ & & 1 \end{pmatrix}.$$

Find a 2×2 matrix that does not commute with $\begin{pmatrix} 1 & 3 \\ 0 & 2 \end{pmatrix}$ For example, we have

$$\begin{pmatrix} 1 & 3 \\ 0 & 2 \end{pmatrix}\begin{pmatrix} 0 & 1 \\ 1 & 0 \end{pmatrix} \neq \begin{pmatrix} 0 & 1 \\ 1 & 0 \end{pmatrix}\begin{pmatrix} 1 & 3 \\ 0 & 2 \end{pmatrix},$$

which means that the scalar multiplication by

$$\begin{pmatrix} 0 & 1 & 0 \\ 1 & 0 & 0 \\ 0 & 0 & 1 \end{pmatrix}$$

does not commute with φ.

[4] We first prove $\ker\varphi = X\,k[X]$. If $f(X) \in \ker\varphi$, then we have $f(0) = 0$. The factor theorem tells us that there is $g(X) \in k[X]$ such that $f(X) = Xg(X)$. This means $f(X) \in X\,k[X]$ and

$$\ker\varphi \subset X\,k[X].$$

Conversely, if $f(X) \in Xk[X]$, then there is $g(X) \in k[X]$ such that $f(X) = Xg(X)$. We have $f(0) = 0$. This means $f(X) \in \ker\varphi$ and

$$X\,k[X] \subset \ker\varphi.$$

We conclude $\ker\varphi = X\,k[X]$.

Next, we prove $\operatorname{Im}\varphi = k$. It is clear that $\operatorname{Im}\varphi \subset k$. We focus on the converse. Take any $a \in k$. Let $f(X) = a \in k[X]$ be a polynomial of degree 0. Then, we have $\varphi(f(X)) = a$ and $k \subset \operatorname{Im}\varphi$. We conclude $\operatorname{Im}\varphi = k$.

[5] The procedure is completely the same as in Exercise [4]. We only present the final results:

$$\begin{aligned} \ker\varphi &= (X-1)\,k[X] = \{(X-1)f(X) \mid f(X) \in k[X]\}, \\ \operatorname{Im}\varphi &= k. \end{aligned}$$

[6] Any $\mathbb{Z}$-submodule of $\mathbb{Z}$ can be expressed as $m\mathbb{Z}$ for some natural number m. If $m\mathbb{Z}$ has a complementary $\mathbb{Z}$-submodule, then it can also be expressed as $n\mathbb{Z}$ for some natural number n. We exclude trivial submodules from our discussion. So, we have $m, n \geq 2$ and

$$mn \in m\mathbb{Z} \cap n\mathbb{Z}.$$

Thus, we have $\mathbb{Z} \neq m\mathbb{Z} \oplus n\mathbb{Z}$ and get a contradiction.

[7] (1) We focus on the closedness under scalar multiplication. For any scalar $\begin{pmatrix} x & y \\ 0 & z \end{pmatrix} \in A$, we have

$$\begin{pmatrix} x & y \\ 0 & z \end{pmatrix} \begin{pmatrix} a & ka \\ 0 & 0 \end{pmatrix} = \begin{pmatrix} xa & xka \\ 0 & 0 \end{pmatrix} = \begin{pmatrix} xa & k(xa) \\ 0 & 0 \end{pmatrix} \in L_k,$$

which shows that L_k is closed under scalar multiplication.

(2) We show that $A = L_k \oplus_A L_\infty$ when $k \neq \infty$. The equation

$$\begin{pmatrix} a & b \\ 0 & c \end{pmatrix} = \begin{pmatrix} a & ka \\ 0 & 0 \end{pmatrix} + \begin{pmatrix} 0 & b - ka \\ 0 & c \end{pmatrix}$$

shows $A = L_k + L_\infty$. If we set

$$\begin{pmatrix} a & ka \\ 0 & 0 \end{pmatrix} = \begin{pmatrix} 0 & b \\ 0 & c \end{pmatrix},$$

then we have $a = b = c = 0$. Thus, we have $L_k \cap L_\infty = \{0_A\}$.

(3) Straightforward calculation.

(4) By (3) we see that

$$\{0_A\}, \quad A, \quad N, \quad \text{and} \quad L_k \ (k \in \mathbb{C} \cup \{\infty\})$$

exhaust left A-submodules of A. None of $\{0_A\}$, A, and N are candidates for a complement of N.

Let us see if $N \cap L_k = \{0_A\}$ and $N + L_k = A$.

Since $N \subsetneq L_\infty \subsetneq A$, the left A-submodule L_∞ is not a complement of N.

If $k < \infty$, then

$$\begin{pmatrix} 0 & b \\ 0 & 0 \end{pmatrix} + \begin{pmatrix} a & ka \\ 0 & 0 \end{pmatrix} = \begin{pmatrix} a & b + ka \\ 0 & 0 \end{pmatrix}$$

shows that $N + L_k \subsetneq A$, which means that the left A-submodule L_k is not a complement of N.

Chapter 10

Quiz 10.1 We calculate blockwise as

$$\left(\begin{array}{c|c} \mathbf{A} & \begin{matrix}0\\0\end{matrix} \\ \hline 0\;0 & a \end{array}\right) \in A \quad \text{and} \quad \left(\begin{array}{c|c} \mathbf{X} & \begin{matrix}0\\0\end{matrix} \\ \hline 0\;0 & 0 \end{array}\right) \in I_1.$$

Left and right scalar multiplications are as follows:

$$\left(\begin{array}{c|c} \mathbf{A} & \begin{matrix}0\\0\end{matrix} \\ \hline 0\;0 & a \end{array}\right)\left(\begin{array}{c|c} \mathbf{X} & \begin{matrix}0\\0\end{matrix} \\ \hline 0\;0 & 0 \end{array}\right) = \left(\begin{array}{c|c} \mathbf{AX} & \begin{matrix}0\\0\end{matrix} \\ \hline 0\;0 & 0 \end{array}\right) \in I_1,$$

$$\left(\begin{array}{c|c} \mathbf{X} & \begin{matrix}0\\0\end{matrix} \\ \hline 0\;0 & 0 \end{array}\right)\left(\begin{array}{c|c} \mathbf{A} & \begin{matrix}0\\0\end{matrix} \\ \hline 0\;0 & a \end{array}\right) = \left(\begin{array}{c|c} \mathbf{XA} & \begin{matrix}0\\0\end{matrix} \\ \hline 0\;0 & 0 \end{array}\right) \in I_1.$$

These results show that I_1 is a two-sided ideal.

For any element

$$\left(\begin{array}{c|c} \mathbf{0} & \begin{matrix}0\\0\end{matrix} \\ \hline 0\;0 & x \end{array}\right) \in I_2,$$

the left and right scalar multiplications are calculated as follows:

$$\left(\begin{array}{c|c} \mathbf{A} & \begin{matrix}0\\0\end{matrix} \\ \hline 0\;0 & a \end{array}\right)\left(\begin{array}{c|c} \mathbf{0} & \begin{matrix}0\\0\end{matrix} \\ \hline 0\;0 & x \end{array}\right) = \left(\begin{array}{c|c} \mathbf{0} & \begin{matrix}0\\0\end{matrix} \\ \hline 0\;0 & ax \end{array}\right) \in I_2,$$

$$\left(\begin{array}{c|c} \mathbf{0} & \begin{matrix}0\\0\end{matrix} \\ \hline 0\;0 & x \end{array}\right)\left(\begin{array}{c|c} \mathbf{A} & \begin{matrix}0\\0\end{matrix} \\ \hline 0\;0 & a \end{array}\right) = \left(\begin{array}{c|c} \mathbf{0} & \begin{matrix}0\\0\end{matrix} \\ \hline 0\;0 & xa \end{array}\right) \in I_2.$$

These results show that I_2 is a two-sided ideal.

Quiz 10.2 Using the techniques introduced in Sect. 1.5, we immediately see that I_3 is closed under the left scalar multiplication and that I_4 is closed under the right scalar multiplication.

The set I_3 is not closed under the right scalar multiplication:

$$\begin{pmatrix} x_{11} & 0 & 0 \\ x_{21} & 0 & 0 \\ 0 & 0 & 0 \end{pmatrix}\begin{pmatrix} 0 & 1 & 0 \\ 1 & 0 & 0 \\ 0 & 0 & 1 \end{pmatrix} = \begin{pmatrix} 0 & x_{11} & 0 \\ 0 & x_{21} & 0 \\ 0 & 0 & 0 \end{pmatrix} \notin I_3.$$

The set I_4 is not closed under the left scalar multiplication:

$$\begin{pmatrix} 0 & 1 & 0 \\ 1 & 0 & 0 \\ 0 & 0 & 1 \end{pmatrix} \begin{pmatrix} x_{11} & x_{12} & 0 \\ 0 & 0 & 0 \\ 0 & 0 & 0 \end{pmatrix} = \begin{pmatrix} 0 & 0 & 0 \\ x_{11} & x_{12} & 0 \\ 0 & 0 & 0 \end{pmatrix} \notin I_4.$$

Quiz 10.3 Let d be the greatest common divisor of a and b, and m be the least common multiple. We have

$$a\mathbb{Z} \cap b\mathbb{Z} = m\mathbb{Z}, \quad a\mathbb{Z} + b\mathbb{Z} = d\mathbb{Z}.$$

Review Sect. 1.3.

Quiz 10.4 The Euclidean division of X^3 by X^3 yields the remainder 0.

Exercises

[1] The submodules of $\mathbb{Z}_{12} = \{0, 1, 2, \cdots, 11\}$ are listed as $2\mathbb{Z}_{12}$, $3\mathbb{Z}_{12}$, $4\mathbb{Z}_{12}$, and $6\mathbb{Z}_{12}$ except for trivial ones and $4\mathbb{Z}_{12} = \{0, 4, 8\}$.

(1) We have

$$4\mathbb{Z}_{12} + 2\mathbb{Z}_{12} = 4\mathbb{Z}_{12},$$
$$4\mathbb{Z}_{12} + 6\mathbb{Z}_{12} = 2\mathbb{Z}_{12},$$

which means the remaining candidate for the complement of $4\mathbb{Z}$ is $3\mathbb{Z}_{12}$ only. The table

$4\mathbb{Z}_{12}$	$3\mathbb{Z}_{12}$	$4\mathbb{Z}_{12} + 3\mathbb{Z}_{12}$
0	0	0
0	3	3
0	6	6
0	9	9
4	0	4
4	3	7
4	6	10
4	9	1
8	0	8
8	3	11
8	6	2
8	9	5

shows that

$$4\mathbb{Z}_{12} + 3\mathbb{Z}_{12} = \mathbb{Z}_{12} \quad \text{and} \quad 4\mathbb{Z}_{12} \cap 3\mathbb{Z}_{12} = \{0\}.$$

So, we conclude $\mathbb{Z}_{12} = 4\mathbb{Z}_{12} \oplus 3\mathbb{Z}_{12}$.

(2) We have the following:

$$2\mathbb{Z}_{12} + 3\mathbb{Z}_{12} = \mathbb{Z}_{12},$$
$$2\mathbb{Z}_{12} + 4\mathbb{Z}_{12} = 2\mathbb{Z}_{12},$$
$$2\mathbb{Z}_{12} + 6\mathbb{Z}_{12} = 2\mathbb{Z}_{12}.$$

At this point, the remaining candidate for the complement of $2\mathbb{Z}$ is $3\mathbb{Z}_{12}$ only. However, we have

$$2\mathbb{Z}_{12} \cap 3\mathbb{Z}_{12} = 6\mathbb{Z}_{12} \neq \{0\},$$

which means $3\mathbb{Z}_{12}$ is not a complement of $2\mathbb{Z}_{12}$.

[2] The map α, "change of variables," defined by

$$\alpha : \mathbb{C}[X, Y] \to \mathbb{C}[X', Y']$$
$$f(X, Y) \mapsto f(X' + 1, Y' + 2)$$

is a ring isomorphism. In fact, substitution preserves addition, multiplication, and the unity, which guarantees that the map α is a ring homomorphism. The inverse homomorphism is easy to write down. Thus, α is a ring isomorphism. We have a ring homomorphism $\varphi' : \mathbb{C}[X', Y'] \to \mathbb{C}$ that makes the following diagram commute:

$$\begin{array}{ccc} \mathbb{C}[X, Y] & \xrightarrow{\alpha} & \mathbb{C}[X', Y'] \\ \varphi \big\downarrow & \circlearrowleft & \big\downarrow \varphi' \\ \mathbb{C} & = & \mathbb{C} \end{array}$$

Just taking $\varphi' = \varphi \circ \alpha^{-1}$, we have the following:

$$\varphi' : \mathbb{C}[X', Y'] \to \mathbb{C},$$
$$f(X', Y') \mapsto f(0, 0).$$

From Example 10.12, we see immediately that

$$\ker \varphi' = X' \,\mathbb{C}[X, Y] + Y' \,\mathbb{C}[X, Y],$$
$$\ker \varphi = (X - 1)\,\mathbb{C}[X, Y] + (Y - 2)\,\mathbb{C}[X, Y].$$

The reader who is not satisfied with the argument by change of variables may take an alternative approach by the Taylor expansions. It is clear that

$$(X-1)\,\mathbb{C}[X,Y]+(Y-2)\,\mathbb{C}[X,Y]\subset\ker\varphi.$$

We prove the converse. For any $f(X,Y)\in\mathbb{C}[X,Y]$, the Taylor expansion around the point $(1,2)$ is expressed as follows:

$$\begin{aligned}f(X,Y)&=c_0+c_{10}(X-1)+c_{01}(Y-2)+c_{20}(X-1)^2\\&\quad+c_{11}(X-1)(Y-2)+c_{02}(Y-2)^2\\&\quad+\cdots\quad\text{(terminates in finite steps).}\end{aligned}$$

Assume $f(1,2)=0$. Then, we have $c_0=0$. If we collect all terms that are free of a factor $X-1$, we have $(Y-2)\,h(Y-2)$ for some polynomial $h(*)$ of one variable. The remainder $f(X,Y)-(Y-2)\,h(Y-2)$ contains all terms that have a factor $X-1$. We have

$$f(X,Y)-(Y-2)\,h(Y-2)=(X-1)\,g(X-1,Y-2)$$

for some polynomial $g(*,*)$. Thus,

$$f(X,Y)=(X-1)\,g(X-1,Y-2)+(Y-2)\,h(Y-2)\in I.$$

[3] The factor ring is expressed as $\mathbb{C}[X]/I=\{aX+b+I\mid a,b\in\mathbb{C}\}$. We write $\overline{aX+b}$ instead of $aX+b+I$ in the following. Since

$$\begin{aligned}(aX+b)(cX+d)&=acX^2+(ad+bc)X+bd\\&=ac(X^2+X+1)+(ad+bc-ac)X+bd-ac,\end{aligned}$$

we have the multiplication formula

$$\overline{aX+b}\cdot\overline{cX+d}=\overline{(ad+bc-ac)X+bd-ac}.\tag{A.1}$$

Since the polynomial X^2+X+1 can be factorized as

$$\left(X-\frac{-1+\sqrt{3}i}{2}\right)\left(X-\frac{-1-\sqrt{3}i}{2}\right),$$

we have

$$\overline{X-\frac{-1+\sqrt{3}i}{2}}\cdot\overline{X-\frac{-1-\sqrt{3}i}{2}}=\bar{0}.$$

This shows the existence of zero-divisors. Thus, $\mathbb{C}[X]/I$ is not an integral domain. The ideal I is not a prime ideal.

[4] The factor ring is expressed as $\mathbb{Q}[X]/I = \{aX + b + I \mid a, b \in \mathbb{Q}\}$. We write $\overline{aX+b}$ instead of $aX + b + I$ in the following. The multiplication formula is exactly the same as Eq. (A.1). Since $X^2 + X + 1$ cannot be factorized over $\mathbb{Q}$, we expect that $\overline{aX+b} \neq \bar{0}$ has a multiplicative inverse. Let c, d be unknowns and consider the system of equations:

$$\begin{cases} ad + bc - ac = 0 \\ bd - ac = 1 \end{cases}.$$

In a matrix form, we have

$$\begin{pmatrix} b-a & a \\ -a & b \end{pmatrix} \begin{pmatrix} c \\ d \end{pmatrix} = \begin{pmatrix} 0 \\ 1 \end{pmatrix}.$$

We want to prove that

$$\det \begin{pmatrix} b-a & a \\ -a & b \end{pmatrix} = a^2 - ab + b^2 \neq 0$$

holds for all $a, b \in \mathbb{Q}$ ($a \neq 0$ or $b \neq 0$). If $b = 0$, then $a \neq 0$ therefore $a^2 - ab + b^2 = a^2 \neq 0$. If $b \neq 0$, then the completed square

$$a^2 - ab + b^2 = \left(a - \frac{b}{2}\right)^2 + \frac{3}{4}b^2$$

shows $a^2 - ab + b^2 \neq 0$.

We conclude that the ring $\mathbb{Q}[X]/I$ is a field and that I is a prime ideal of $\mathbb{Q}[X]$.

[5] Substitution is an easy way of creating ring homomorphisms.

(1) We define a ring homomorphism $\varphi : \mathbb{C}[X, Y, Z] \to \mathbb{C}[Y, Z]$ by

$$\varphi : f(X, Y, Z) \mapsto f(0, Y, Z).$$

If $f(0, Y, Z) = 0$, then regarding Y, Z as constants and applying the factor theorem, we have

$$f(X, Y, Z) = (X - 0)g(X, Y, Z),$$

which shows that $\ker\varphi = I_1$ and thereby $\mathbb{C}[X, Y, Z]/I_1 \simeq \mathbb{C}[Y, Z]$.

(2) We define a ring homomorphism $\varphi : \mathbb{C}[X, Y, Z] \to \mathbb{C}[Z]$ by

$$\varphi : f(X, Y, Z) \mapsto f(0, 0, Z).$$

It is clear that

$$I_2 = X\,\mathbb{C}[X, Y, Z] + Y\,\mathbb{C}[X, Y, Z] \subset \ker\varphi.$$

We prove the inclusion reversed. Regarding Z as a constant, we have the Taylor expansion

$$\begin{aligned} f(X, Y, Z) =& c_{00} + c_{10}X + c_{01}Y + c_{20}X^2 + c_{11}XY + c_{02}Y^2 \\ &+ \cdots \quad \text{(terminates in finite steps)} \end{aligned}$$

around $(X, Y) = (0, 0)$, where $c_{ij} \in \mathbb{C}[Z]$. If $f(0, 0, Z) = 0$, then $c_{00} = 0$. All terms free of X have the form $c_{0n}Y^n \quad (n \geq 1)$, the sum of which can be expressed as $Y\,h(Y, Z)$ for some polynomial $h(Y, Z)$. The expression $f(X, Y, Z) - Y\,h(Y, Z)$ is the sum of all terms with X as a factor. Thus, $f(X, Y, Z) - Y\,h(Y, Z) = X\,g(X, Y, Z)$ for some polynomial $g(X, Y, Z)$. To sum up, any $f(X, Y, Z) \in \ker\varphi$ can be expressed as

$$f(X, Y, Z) = X\,g(X, Y, Z) + Y\,h(Y, Z),$$

which shows

$$\ker\varphi \subset X\,\mathbb{C}[X, Y, Z] + Y\,\mathbb{C}[X, Y, Z].$$

We conclude

$$\begin{aligned} \ker\varphi &= X\,\mathbb{C}[X, Y, Z] + Y\,\mathbb{C}[X, Y, Z] \\ &\text{and thereby} \quad \mathbb{C}[X, Y, Z]/I_2 \simeq \mathbb{C}[Z]. \end{aligned}$$

(3) We define a ring homomorphism $\varphi : \mathbb{C}[X, Y, Z] \to \mathbb{C}$ by

$$\varphi : f(X, Y, Z) \mapsto f(0, 0, 0).$$

It is clear that

$$I_3 = X\,\mathbb{C}[X, Y, Z] + Y\,\mathbb{C}[X, Y, Z] + Z\,\mathbb{C}[X, Y, Z] \subset \ker\varphi.$$

We prove the inclusion reversed. We have the Taylor expansion

$$\begin{aligned} f(X, Y, Z) = & c_{000} + c_{100}X + c_{010}Y + c_{001}Z \\ &+ c_{200}X^2 + c_{110}XY + c_{101}XZ \\ &+ c_{020}Y^2 + c_{011}YZ + c_{002}Z^2 \\ &+ \cdots \quad \text{(terminates in finite steps)} \end{aligned}$$

around $(X, Y, Z) = (0, 0, 0)$, where $c_{ijk} \in \mathbb{C}$. If $f(0, 0, 0) = 0$, then $c_{000} = 0$. All terms free of X, Y have the form $c_{00n}Z^n$ $(n \geq 1)$, the sum of which can be expressed as $Z\,r(Z)$. The expression $f(X, Y, Z) - Z\,r(Z)$ consists of terms with X or Y as a factor. Among them, the sum of the terms free of X is expressed as $Y\,q(Y, Z)$ for some polynomial $q(Y, Z)$. The expression $f(X, Y, Z) - Z\,r(Z) - Y\,q(Y, Z)$ has terms with X as a factor. Therefore,

$$f(X, Y, Z) - Z\,r(Z) - Y\,q(Y, Z) = X\,p(X, Y, Z)$$

for some polynomial $p(X, Y, Z)$.

To sum up, any $f(X, Y, Z) \in \ker\varphi$ can be expressed as

$$f(X, Y, Z) = X\,p(X, Y, Z) + Y\,q(Y, Z) + Z\,r(Z),$$

which shows

$$\ker\varphi \subset X\,\mathbb{C}[X, Y, Z] + Y\,\mathbb{C}[X, Y, Z] + Z\,\mathbb{C}[X, Y, Z].$$

We conclude

$$\ker\varphi = X\,\mathbb{C}[X, Y, Z] + Y\,\mathbb{C}[X, Y, Z] + Z\,\mathbb{C}[X, Y, Z]$$
$$\text{and thereby} \quad \mathbb{C}[X, Y, Z]/I_3 \simeq \mathbb{C}.$$

[6]

$$\begin{array}{ccc}
\mathbb{Z} & & \mathbb{Z}/(8\mathbb{Z}) \\
\cup & & \cup \\
2\mathbb{Z} & \longmapsto & 2(\mathbb{Z}/(8\mathbb{Z})) = \{\bar{0}, \bar{2}, \bar{4}, \bar{6}\} \\
\cup & & \cup \\
4\mathbb{Z} & \longmapsto & 4(\mathbb{Z}/(8\mathbb{Z})) = \{\bar{0}, \bar{4}\} \\
\cup & & \cup \\
8\mathbb{Z} & \longmapsto & \{\bar{0}\}
\end{array}$$

[7] (1) $\{0\}$. (2) $Y B$.

[8] Let m_1, m_2, m_3, M_1, M_2, and M_3 be as follows:

$$m_1 = 2^3, \quad m_2 = 3^2, \quad m_3 = 5,$$
$$M_1 = 3^2 \times 5, \quad M_2 = 2^3 \times 5, \quad M_3 = 2^3 \times 3^2.$$

By Euclidean algorithm, we have the following:

$$17m_1 - 3M_1 = 1,$$
$$9m_2 - 2M_2 = 1,$$
$$29m_3 - 2M_3 = 1.$$

We have

$$\varphi^{-1} :(u + 8\mathbb{Z}, v + 9\mathbb{Z}, w + 5\mathbb{Z}) \mapsto (u \times (-3) \times 45 + v \times (-2) \times 40 + w \times (-2) \times 72) + 360\mathbb{Z}.$$

Chapter 11

Exercises

[1] (1) $\mathbb{F}_5^\times = \{\bar{1}, \bar{2}, \bar{3}, \bar{4}\}$ is cyclic.

×	1	2	3	4	inverse	order
1	1	2	3	4	1	1
2	2	4	1	3	3	4
3	3	1	4	2	2	4
4	4	3	2	1	4	2

(2) $\mathbb{F}_7^\times = \{\bar{1}, \bar{2}, \bar{3}, \bar{4}, \bar{5}, \bar{6}\}$ is cyclic.

×	1	2	3	4	5	6	inverse	order
1	1	2	3	4	5	6	1	1
2	2	4	6	1	3	5	4	3
3	3	6	2	5	1	4	5	6
4	4	1	5	2	6	3	2	3
5	5	3	1	6	4	2	3	6
6	6	5	4	3	2	1	6	2

(3) $\mathbb{F}_{11}^\times = \{\bar{1}, \bar{2}, \ldots, \overline{10}\}$ is cyclic.

×	1	2	3	4	5	6	7	8	9	10	inverse	order
1	1	2	3	4	5	6	7	8	9	10	1	1
2	2	4	6	8	10	1	3	5	7	9	6	10
3	3	6	9	1	4	7	10	2	5	8	4	5
4	4	8	1	5	9	2	6	10	3	7	3	5
5	5	10	4	9	3	8	2	7	1	6	9	5
6	6	1	7	2	8	3	9	4	10	5	2	10
7	7	3	10	6	2	9	5	1	8	4	8	10
8	8	5	2	10	7	4	1	9	6	3	7	10
9	9	7	5	3	1	10	8	6	4	2	5	5
10	10	9	8	7	6	5	4	3	2	1	10	2

[2] (1) $(\mathbb{Z}/15\mathbb{Z})^\times = \{\bar{1}, \bar{2}, \bar{4}, \bar{7}, \bar{8}, \overline{11}, \overline{13}, \overline{14}\}$

×	1	2	4	7	8	11	13	14	inverse	order
1	1	2	4	7	8	11	13	14	1	1
2	2	4	8	14	1	7	11	13	8	4
4	4	8	1	13	2	14	7	11	4	2
7	7	14	13	4	11	2	1	8	13	4
8	8	1	2	11	4	13	14	7	2	4
11	11	7	14	2	13	1	8	4	11	2
13	13	11	7	1	14	8	4	2	7	4
14	14	13	11	8	7	4	2	1	14	2

Let us choose $\bar{2}$ as an element of order 4. We have

$$\bar{2}^2 = \bar{4}, \quad \bar{2}^3 = \bar{8}, \quad \bar{2}^4 = \bar{1}.$$

Let us choose $\overline{11}$ as an element of order 2 that is not included in $\{\bar{1}, \bar{2}, \bar{2}^2, \bar{2}^3\}$. Then, we have

$$(\mathbb{Z}/15\mathbb{Z})^\times \cong \{\bar{1}, \bar{2}, \bar{2}^2, \bar{2}^3\} \times \{\bar{1}, \overline{11}\}.$$

Other choices and combinations are possible.

(2) $(\mathbb{Z}/8\mathbb{Z})^\times = \{\bar{1}, \bar{3}, \bar{5}, \bar{7}\}$

×	1	3	5	7	inverse	order
1	1	3	5	7	1	1
3	3	1	7	5	3	2
5	5	7	1	3	5	2
7	7	5	3	1	7	2

We see $(\mathbb{Z}/8\mathbb{Z})^\times = K_4$.

[3] (1) $(\mathbb{F}_2[X]/I_1)^\times$ is cyclic.

$\times$	1	X	$X+1$	X^2	X^2+1	X^2+X	X^2+X+1	order
1	1	X	$X+1$	X^2	X^2+1	X^2+X	X^2+X+1	1
X	X	X^2	X^2+X	$X+1$	1	X^2+X+1	X^2+1	7
$X+1$	$X+1$	X^2+X	X^2+1	X^2+X+1	X^2	1	X	7
X^2	X^2	$X+1$	X^2+X+1	X^2+X	X	X^2+1	1	7
X^2+1	X^2+1	1	X^2	X	X^2+X+1	$X+1$	X^2+X	7
X^2+X	X^2+X	X^2+X+1	1	X^2+1	$X+1$	X	X^2	7
X^2+X+1	X^2+X+1	X^2+1	X	1	X^2+X	X^2	$X+1$	7

This table is interesting in itself. However, without this table, it is easy to see that $\mathbb{F}_2[X]/I_1$ is a field. Set

$$f(X) = X^3 + X + 1.$$

Since $f(0) = 1$ and $f(1) = 1$, the polynomial $f(X)$ is irreducible. Therefore, the ring $\mathbb{F}_2[X]/I_1$ is a field.

The polynomial $X^3 + X^2 + 1$ is also irreducible. Therefore, the ring $\mathbb{F}_2[X]/I_2$ is a field. The table is given below. The multiplicative group $(\mathbb{F}_2[X]/I_2)^\times$ is cyclic.

$\times$	1	X	$X+1$	X^2	X^2+1	X^2+X	X^2+X+1	order
1	1	X	$X+1$	X^2	X^2+1	X^2+X	X^2+X+1	1
X	X	X^2	X^2+X	X^2+1	X^2+X+1	1	$X+1$	7
$X+1$	$X+1$	X^2+X	X^2+1	1	X	X^2+X+1	X^2	7
X^2	X^2	X^2+1	1	X^2+X+1	$X+1$	X	X^2+X	7
X^2+1	X^2+1	X^2+X+1	X	$X+1$	X^2+X	X^2	1	7
X^2+X	X^2+X	1	X^2+X+1	X	X^2	$X+1$	X^2+1	7
X^2+X+1	X^2+X+1	$X+1$	X^2	X^2+X	1	X^2+1	X	7

(2) We express elements of $\mathbb{F}_2[X]/I_1$ with bars and elements of $\mathbb{F}_2[X]/I_2$ with double bars. Set $f(X) = X^3 + X + 1$. We seek a solution of the equation $f(X) = 0$ in the field $\mathbb{F}_2[X]/I_2$. We line up all elements of $\mathbb{F}_2[X]/I_2$ in lexicographical order and try substitution one by one as follows:

$$f(\bar{\bar{0}}) = \bar{\bar{1}},$$

$$f(\bar{\bar{X}}) = \overline{\overline{X^2 + X}},$$

$$f(\overline{\overline{X+1}}) = \bar{\bar{0}}.$$

We have found one.

Noting that $\mathbb{F}_2[X] = \mathbb{F}_2[X+1]$, we have the following:

$$\begin{aligned} &\mathbb{F}_2[X]/(X^3 + X + 1)\mathbb{F}_2[X] \\ \cong\ &\mathbb{F}_2[X+1]/((X+1)^3 + (X+1) + 1)\mathbb{F}_2[X+1] \\ =\ &\mathbb{F}_2[X]/((X+1)^3 + (X+1) + 1)\mathbb{F}_2[X] \end{aligned}$$

$$= \mathbb{F}_2[X]/(X^3 + X^2 + 1)\mathbb{F}_2[X].$$

We can construct a field isomorphism $\varphi : \mathbb{F}_2[X]/I_1 \to \mathbb{F}_2[X]/I_2$ by setting

$$\varphi : \overline{X} \mapsto \overline{\overline{X+1}},$$

which yields uniquely the whole correspondence

$$\varphi : a\overline{X}^2 + b\overline{X} + c \mapsto a\overline{\overline{X+1}}^2 + b\overline{\overline{X+1}} + c.$$

[4] Set $f(X) = X^2 + X + 1$. Since $f(1) = 1 + 1 + 1 = 0$, we see that $f(X)$ has a divisor $X - 1 = X + 2$. Recall that $-1 = 2$ in $\mathbb{F}_3$. Divide $f(X)$ by $X + 2$ as follows:

$$\begin{array}{r|l} & X + 2 \\ \hline X+2 & X^2 + X + 1 \\ & X^2 + 2X \\ \hline & \quad\; 2X + 1 \\ & \quad\; 2X + 1 \\ \hline & \qquad\quad 0 \end{array}$$

Thus, we have $X^2 + X + 1 = (X + 2)^2$.

The zero-divisors of $\mathbb{F}_3[X]/I$ are $\overline{X+2}$ and its multiple $\overline{X+2}$. The table of $(\mathbb{F}_3[X]/I)^{\times}$ is given below.

	1	2	X	$X+1$	$2X$	$2X+2$	inverse	order
1	1	2	X	$X+1$	$2X$	$2X+2$	1	1
2	2	1	$2X$	$2X+2$	X	$X+1$	2	2
X	X	$2X$	$2X+2$	2	$X+1$	1	$2X+2$	3
$X+1$	$X+1$	$2X+2$	2	X	1	$2X$	$2X$	6
$2X$	$2X$	X	$X+1$	1	$2X+2$	2	$X+1$	6
$2X+2$	$2X+2$	$X+1$	1	$2X$	2	X	X	3

Chapter 12

Quiz 12.1 If we solve the system of equations

$$\begin{cases} x + y + z = 0 \\ z = 0 \end{cases},$$

then we have

$$(x, y, z) = (0, 0, 0) \quad \text{or} \quad (1, 1, 0).$$

The intersection line consists of these two points.
Quiz 12.2 Let $d_i : \mathbb{F}_2 \to \mathbb{Z}$ be a function defined by

$$d_i(x, y) = \begin{cases} 0 & \text{if } x_i = y_i, \\ 1 & \text{if } x_i \neq y_i. \end{cases}$$

The distance function $d : \mathbb{F}_2 \to \mathbb{Z}$ is expressed as

$$d(x, y) = \sum_{i=1}^{n} d_i(x, y). \tag{A.2}$$

Suppose $d(x, y) = 0$. Since $d_i(x, y) \geq 0$ for all i, all of them must be 0. This means $x = y$; condition (1) is satisfied.

As to condition (2), it is sufficient to prove bitwise, namely

$$d_i(x, y) + d_i(y, z) \geq d_i(x, z)$$

for each i.

When $x_i = z_i$, the RHS is 0. So, the (A.2) holds whatever y_i maybe.

When $x_i \neq z_i$, since $\{x_i, y_i, z_i\} \subset \{0, 1\}$, we must have $y_i = x_i$ or $y_i = z_i$, exclusively. We have

$$d_i(x, y) + d_i(y, z) = 1,$$

consequently (A.2).
Quiz 12.3 Suppose we have

$$a_0\vec{e_0} + a_1\vec{e_1} + a_2\vec{e_2} = \vec{0}$$

for some $a_0, a_1, a_2 \in \mathbb{F}$. The LHS is

$$a_2\overline{X^6} + (a_2 + a_1)\overline{X^5} + (a_2 + a_1 + a_0)\overline{X^4} + \cdots .$$

Since $\overline{X^6}, \overline{X^5}, \cdots, \bar{1}$ are linearly independent, we have the following system of equations:

$$\begin{cases} a_2 = 0 \\ a_2 + a_1 = 0 \\ a_2 + a_1 + a_0 = 0 \\ \cdots \end{cases}$$

The system has only one solution, $a_2 = a_1 = a_0 = 0$.
Quiz 12.4 Straightforward calculation.

Exercises

[1] (1) Any plane can be defined uniquely by an equation like

$$ax + by + cz = d$$

up to multiplication by $\bar{2} \in \mathbb{F}_3^\times$. Roughly the planes are classified into three cases:

- $x + by + cz =$ 0 or 1 or 2,
- $y + cz =$ 0 or 1 or 2,
- $z =$ 0 or 1 or 2.

Using this classification, we know there are $(9 + 3 + 1) \times 3 = 39$ planes in total.

We present an alternative solution. The total number of the "normal" vectors, or the coefficient vectors, (a, b, c) of planes is $(3^3 - 1)/2 = 13$, where:

- 3^3 is the number of $\{0, 1, 2\}^3$, the naive choice of a, b, and c,
- $3^3 - 1$ is the number of the above excluding meaningless $(0, 0, 0)$, and
- $/2$ is the adjustment for the freedom of multiplication by $\bar{2} \in \mathbb{F}_3$.

For each "normal vector", there are three parallel planes: the RHS constants can take values from $\{\bar{0}, \bar{1}, \bar{2}\}$. Therefore, we get the same conclusion as the first solution.

(2) We know that the number of the directions (a, b, c) of straight lines (a, b, c) is 13 since the calculation is exactly the same as the alternative solution to (1). Any straight line with direction (a, b, c) meets the plane $ax+by+cz = t$ at one point. Recall that every plane has $3 \times 3 = 9$ points. Therefore, the number of straight lines is $13 \times 9 = 117$.

We present an alternative solution. A straight line is determined when two different points on it are given. Therefore, the number of straight lines is ${}_{27}C_2/{}_3C_2 = 117$, where the denominator ${}_3C_2$ is the adjustment for the duplication that any straight line has 3 points on it.

[2] (1) The equation of a plane is

$$a_1x_1 + a_2x_2 + a_3x_3 + a_4x_4 = 0 \quad \text{or} \quad 1$$

for some $a_1, a_2, a_3, a_4 \in \mathbb{F}_2$. The number of planes is $(2^4 - 1) \times 2 = 30$.

(2) Any straight line consists of two points. The space ${\mathbb{F}_2}^4$ has a total of $2^4 = 16$ points. So, the number of straight lines in ${\mathbb{F}_2}^4$ is $\binom{16}{2} = 120$.

[3] (1) X and $X + 1$.

(2) A quadratic polynomial is expressed as $X^2 + aX + b$. If $b = 0$, then the polynomial has the factor X. So, we may fix $b = 1$.

We have to check two cases: $X^2 + 1$ and $X^2 + X + 1$. The former is factorized as $(X + 1)^2$.

To study the latter, we set $f(X) = X^2+X+1$. We have $f(0) = f(1) = 1$. This shows that $f(X)$ cannot be factorized in $\mathbb{F}_2[X]$. There is only one irreducible quadratic polynomial

$$X^2 + X + 1.$$

(3) Cubic polynomials are factorized in $\mathbb{F}_2[X]$, if possible, in two ways:

$$(X + a)(X + b)(X + c) \quad \text{or} \quad (X + a)(\text{irreducible quadratic factor}).$$

In either case, such a polynomial must have a linear factor. Therefore, $f(0) = f(1) = 1$ is a necessary and sufficient condition for a cubic polynomial $f(X)$ to be irreducible. Based on this fact, we seek irreducible cubic polynomials. We set

$$f(X) = X^3 + aX^2 + bX + 1,$$

where $a, b \in \mathbb{F}_2$. So, we only have to check $2 \times 2 = 4$ cases. Eventually, we have two irreducible polynomials

$$X^3 + X + 1 \quad \text{and} \quad X^3 + X^2 + 1.$$

(4) Quartic polynomials are factorized in $\mathbb{F}_2[X]$, if possible, in four ways:

$$(X + a)(X + b)(X + c)(X + d),$$
$$(X + a)(X + b)(\text{irreducible quadratic factor}),$$
$$(X + a)(\text{irreducible cubic factor}), \quad \text{and}$$
$$(\text{irreducible quadratic factor})(\text{irreducible quadratic factor}).$$

We seek irreducible quartic polynomials. Since $f(0) \neq 0$, we set

$$f(X) = X^4 + aX^3 + bX^2 + cX + 1.$$

We must have $f(1) = 1$. Therefore, we have $a+b+c = 1$. The possibilities are as follows:

$$(a, b, c) = (1, 1, 1), \quad (1, 0, 0), \quad (0, 1, 0), \quad (0, 0, 1).$$

At this point, we have $f(0) = f(1) = 1$. So, we have no chance of having a linear factor. The only possibility of

$$(\text{irreducible quadratic factor})(\text{irreducible quadratic factor})$$

is as follows:

$$(X^2 + X + 1)^2 = X^4 + X^2 + 1.$$

To sum up, irreducible quartic polynomials are

$$X^4 + X^3 + X^2 + X + 1, \quad X^4 + X^3 + 1, \quad \text{and} \quad X^4 + X + 1.$$

[4] Solve the equation

$$\overline{X^6 + X^4 + 1} = a\vec{f}_3 + b\vec{f}_2 + c\vec{f}_1 + d\vec{f}_0$$

with unknowns a, b, c, d.
Solve other similar equations.

References

There are so many well-established textbooks in algebra. The reader can go to the nearest library and find the right textbook to read next.

I suggest Milne's book (Milne 2021) for group theory. It provides us with a gentle bridge from this book to advanced texts, spanning only 131 pages. It serves as a standard textbook for undergraduates and first-year graduate students.

For the theory of groups, rings, and modules all-inclusive, I suggest , a well-tested textbook from Dover classics, Auslander and Buchbaum's book (Auslander and Buchsbaum 2014). I also recommend Facchini's book (Facchini 2025) from the latest publications. It features the category theoretic viewpoint, a modern style of presentation. If the reader is comfortable with French, I recommend Perrin's book (Perrin 2008). It includes the topics of linear groups, bilinear forms, and orthogonal groups. It also provides us with loads of vivid concrete examples.

If the reader wants to learn more on finite fields and polynomial rings over finite fields, a concise textbook is Mullen and Mumert's book (Mullen 2007). It includes applications to coding theory.

If the reader wants a textbook on coding theory, I recommend Hill's book (Hill 1980). The required knowledge is only linear algebra. An alternative or subsequent text will be Goppa's book (Goppa 1988). It guides the reader to the interplay of coding theory and algebraic geometry.

In the following, I list textbooks with distinctive features, yet fully acceptable as standard textbooks.

Chapters 8 to 11 are intended as a preparation for number theory and algebraic geometry. The next choice of reading is Ono's book (Ono 1994). It provides us with introductory materials for groups, rings, and fields, and brings us to middle-level topics in number theory.

If the reader wants to enjoy algebra over coffee (hot chocolate, alcohol or whatever makes the reader relaxed), I recommend Kuga's book (Kuga 1993). The title is extremely attractive, "Galois's Dream." Évariste Galois would have done further advanced work in topology and differential equations if he had lived longer.

S. Yukita, *First Lectures in Algebra*, Springer Asia Pacific Mathematics Series 10,
https://doi.org/10.1007/978-981-95-5457-7

It presents introductions to topology, group theory, Galois theory, and Fuchsian differential equations. It was based on the lectures delivered by Professor Kuga for first-year students at Tokyo University in the 1960s.

Sergei Duzhin, my old friend from Steklov Institute, a musician and topologist, left an extremely *heartwarming* (math!) book on group theory (Duzhin and Chebotarevsky 2004). The reader will find how group theory works in solving differential equations. Lie's method is concretely presented.

Finally, I introduce my own work, an introductory textbook on category theory and functional programming (Yukita 2024), which shares the same features with this book: full of pictures for understanding abstract concepts and a special emphasis on experiences in calculations.

References

Auslander, M., Buchsbaum, D.: Groups, Rings, Modules. Dover, Mineola (2014)

Duzhin, S.V., Chebotarevsky, B.D.: Transformation Groups for Beginners. American Mathematical Society, Providence (2004)

Facchini, A.: Introduction to Ring and Module Theory. Birkhäuser, Basel (2025)

Goppa, V.D.: Geometry and Codes. Springer, Berlin (1988)

Hill, R.: A First Course in Coding Theory. Oxford University Press, Oxford (1980)

Kuga, M.: Galois's Dream: Group Theory and Differential Equations. Birkhäuser, Basel (1993)

Milne, J.S.: Group theory (v4.00) (2021). http://creativecommons.org. Available at www.jmilne.org/math/

Mullen, G.L., Mummert, C.: Finite Fields and Application. Student Mathematical Library. American Mathematical Society, Providence (2007)

Ono, T.: Variations on a Theme of Euler: Quadratic Forms, Elliptic Curves, and Hopf Maps. Springer, Berlin (1994)

Perrin, D.: Cours d'algèbre. Ellipses (2008)

Yukita, S.: Category Theory Using Haskell. Birkhäuser, Basel (2024)

Index

Symbols

A

B

S. Yukita, *First Lectures in Algebra*, Springer Asia Pacific Mathematics Series 10,
https://doi.org/10.1007/978-981-95-5457-7

MIX
Papier aus verantwortungsvollen Quellen
Paper from responsible sources
FSC® C105338

If you have any concerns about our products,
you can contact us on
ProductSafety@springernature.com

In case Publisher is established outside the EU,
the EU authorized representative is:
Springer Nature Customer Service Center GmbH
Europaplatz 3, 69115 Heidelberg, Germany

Printed by Libri Plureos GmbH
in Hamburg, Germany